Lecture Notes
in Control and Information Sciences 251

Editors: M. Thoma · M. Morari

A.J. van der Schaft
J.M. Schumacher

An Introduction to
Hybrid Dynamical Systems

 Springer

Authors

A.J. (Arjan) van der Schaft
Institute for Mathematics and Computer Science
University of Groningen
The Netherlands

J.M. (Hans) Schumacher
Department of Econometrics and Operations Research
Tilburg University
The Netherlands

British Library Cataloguing in Publication Data
A catalogue record for this book is available from the British Library

Library of Congress Control Number: 0000000

Lecture Notes in Control and Information Sciences ISSN 0170-8643

ISBN 978-1-4471-3916-4   ISBN 978-1-84628-542-4 (eBook)

DOI 10.1007/978-1-84628-542-4

Printed on acid-free paper

Preface

The term "hybrid system" has many meanings, one of which is: a dynamical system whose evolution depends on a coupling between variables that take values in a continuum and variables that take values in a finite or countable set. For a typical example of a hybrid system in this sense, consider a temperature control system consisting of a heater and a thermostat. Variables that would in all likelihood be included in a model of such a system are the room temperature and the operating mode of the heater (on or off). It is natural to model the first variable as real-valued and the second as Boolean. Obviously, for the temperature control system to be effective there needs to be a coupling between the continuous and discrete variables, so that for instance the operating mode will be switched to on if the room temperature decreases below a certain value.

Actually most of the dynamical systems that we have around us may reasonably be described in hybrid terms: cars, computers, airplanes, washing machines—there is no lack of examples. Nevertheless, most of the literature on dynamic modeling is concerned with systems that are either completely continuous or completely discrete. There are good reasons for choosing a description in either the continuous or the discrete domain. Indeed, it is a platitude that it is not necessary or even advisable to include all aspects of a given physical system into a model that is intended to answer certain types of questions. The engineering solution to a hybrid system problem therefore has often been to look for a formulation that is primarily continuous or discrete, and to deal with aspects from the other domain, if necessary, in an *ad hoc* manner. As a consequence, the field of hybrid system modeling has been dominated by patches and workarounds.

Indications are, however, that the interaction between discrete and continuous systems in today's technological problems has become so important that more systematic ways of dealing with hybrid systems are called for. For a dramatic example, consider the loss of the Ariane 5 launcher that went into self-destruction mode 37 seconds after liftoff on June 4, 1996. Investigators have put the blame for the costly failure on a software error. Nevertheless, the program that went astray was the same as the one that had worked perfectly in Ariane 4; in fact, it was copied from Ariane 4 to Ariane 5 for exactly that reason. What had changed was the continuous dynamical system around the software, embodied in the physical structure of the new launcher which had been sized up considerably compared to its predecessor. Within the new physical environment, the trusted code quickly led into a catastrophe.

Although the increasing role of the computer in the control of physical processes may be cited as one of the reasons for the increased interest in hybrid systems, there are also other sources of inspiration. In fact a number of recent developments all revolve in some way around the combination of continuous and discrete aspects. The following is a sample of these developments, which are connected in ways that are still largely unexplored:

- computer science: verification of correctness of programs interacting with continuous environments (embedded systems);

- control theory: hierarchical control, interaction of data streams and physical processes, stabilization of nonlinear systems by switching control;

- dynamical systems: discontinuous systems show new types of bifurcations and provide relatively tractable examples of chaos;

- mathematical programming: optimization and equilibrium problems with inequality constraints can fruitfully be placed within a regime-switching dynamic framework;

- simulation languages: element libraries contain both continuous and discrete elements, so that the numerical simulation routines behind the languages must take both aspects into account.

It is a major challenge to advance and systematize the knowledge about hybrid systems that comes from such a large variety of fields, which nevertheless from a historical point of view have many concepts in common.

Even though the overall area of hybrid systems has not yet crystallized, we believe that it is meaningful at this time to take stock of a number of related developments in an introductory text, and to describe these from a more unified point of view. We have not tried to be encyclopaedic, and in any case we consider the present text as an intermediate product. Our own background is clearly reflected in the choice of the developments covered, and without doubt the reader will recognize a definite emphasis on aspects that are of interest from the point of view of continuous dynamics and mathematical systems theory. The title that we have chosen is intended to reflect this choice. We trust that others who are more qualified to do so will write books on hybrid systems emphasizing different aspects.

This text is an expanded and revised version of course notes that we have written for the course in Hybrid Systems that we taught in the spring of 1998 as part of the national graduate course program of the Dutch Institute of Systems and Control. We would like to thank the board of DISC for giving us the opportunity to present this course, and we are grateful to the course participants for their comments. In some parts of the book we have relied heavily on work that we have done jointly with Kanat Çamlıbel, Gerardo Escobar,

Maurice Heemels, Jun-Ichi Imura, Yvonne Lootsma, Romeo Ortega, and Siep Weiland; it is a pleasure to acknowledge their contributions. The second author would in particular like to thank Gjerrit Meinsma for his patient guidance on the intricacies of LaTeX. Finally, for their comments on preliminary versions of this text and helpful remarks, we would like to thank René Boel, Peter Breedveld, Ed Brinksma, Bernard Brogliato, Domine Leenaerts, John Lygeros, Oded Maler, Sven-Erik Mattson, Gjerrit Meinsma, Manuel Monteiro Marques, Andrew Paice, Shankar Sastry, David Stewart, and Jan Willems. Of course, all remaining faults and fallacies are entirely our own.

Enschede/Amsterdam, September 1999

Arjan van der Schaft
Hans Schumacher

On the occasion of the book's transition to print-on-demand status, a few corrections have been made. We would like to thank Kanat Çamlıbel, Bart De Schutter, and anonymous reviewers of papers related to the book for pointing out some errors. The errors that remain are still our responsibility.

Groningen/Tilburg, January 2007

Arjan van der Schaft
Hans Schumacher

Contents

List of Figures

Chapter 1

Modeling of hybrid systems

1.1   Introduction

The aim of this chapter is to make more precise what we want to understand by a *"hybrid system"*. This will be done in a somewhat tentative manner, without actually ending up with a single final definition of a hybrid system. Partly, this is due to the fact that the area of hybrid systems is still in its infancy and that a general theory of hybrid systems seems premature. More inherently, hybrid systems is such a wide notion that sticking to a single definition shall be too restrictive (at least at this moment) for our purposes. Moreover, the choice of hybrid models crucially depends on their purpose, e.g. for theoretical purposes or for specification purposes, or as a simulation language. Nevertheless, we hope to make reasonably clear what should be the main ingredients in any definition of a hybrid system, by proposing and discussing in Section 1.2 a number of definitions of hybrid systems. Subsequently, Chapter 2 will present a series of examples of hybrid systems illustrating the main ingredients of these definitions in more detail.

Generally speaking, hybrid systems are mixtures of real-time (continuous) dynamics and discrete events. These continuous and discrete dynamics not only coexist, but *interact* and changes occur both in response to discrete, instantaneous, events and in response to dynamics as described by differential or difference equations in time.

One of the main difficulties in the discussion of hybrid systems is that the term "hybrid" is not restrictive—the interpretation of the term could be stretched to include virtually any dynamical system we can think of. A reasonably general definition of hybrid systems can therefore only serve as a *framework*, to indicate the main issues and to fix the terminology. *Within* such a general framework one necessarily has to restrict to special subclasses of hybrid systems in order to derive useful generally valid propositions.

Another difficulty in discussing hybrid systems is that various scientific communities with their own approaches have contributed (and are still contributing) to the area. At least the following three communities can be distinguished.

First there is the *computer science* community that looks at a hybrid system primarily as a discrete (computer) program interacting with an analog

1

environment. (In this context also the terminology *embedded systems* is being used.) A leading objective is to extend standard program analysis techniques to systems which incorporate some kind of continuous dynamics. The emphasis is often on the discrete event dynamics, whereas the continuous dynamics is frequently of a relatively simple form. One of the key issues is *verification*.

Another community involved in the study of hybrid systems is the *modeling and simulation* community. Physical systems can often operate in different *modes*, and the transition from one mode to another sometimes can be idealized as an instantaneous, discrete, transition. Examples include electrical circuits with switching devices such as (ideal) diodes and transistors, and mechanical systems subject to inequality constraints as encountered e.g. in robotics. Since the time scale of the transition from one mode to another is often much faster than the time scale of the dynamics of the individual modes, it may be advantageous to model the transitions as being instantaneous. The time instant at which the transition takes place is called an *event time*. Basic issues then concern the well-posedness of the resulting hybrid system, e.g. the existence and uniqueness of solutions, and the ability to efficiently *simulate* the multi-modal physical system.

Yet another community contributing to the area of hybrid systems is the *systems and control* community. Within this community additional motivation for the study of hybrid systems is actually provided from different angles. One can think of hierarchical systems with a discrete decision layer and a continuous implementation layer (e.g. supervisory control or multi-agent control). Also *switching control* schemes and *relay control* immediately lead to hybrid systems. For *nonlinear control* systems it is known that in some important cases there does not exist a *continuous* stabilizing state feedback, but that nevertheless the system can be stabilized by a switching control. Finally, *discrete event systems theory* can be seen as a special case of hybrid systems theory. In many areas of control, e.g. in power converters and in motion control, control strategies are inherently hybrid in nature.

From a general system-theoretic point of view one can look at hybrid systems as systems having two different types of ports through which they interact with their environment. One type of ports consists of the *communication ports*. The variables associated with these ports are *symbolic* in nature, and represent "data flow". The strings of symbols at these communication ports in general are not directly related with real (physical) time; there is only a sequential ordering. The second type of ports consists of the *physical ports*, where the term "physical" interpreted in a broad sense; perhaps "analog" would be a more appropriate terminology. The variables at these ports are usually continuous variables, and are related to physical measurement. Also, the flow of these variables is directly related to physical time. In principle the signals at the physical ports may be discrete-time signals (or sampled-data signals), but in most cases they will ultimately be *continuous-time* signals.

Thus a hybrid system can be regarded as a combination of discrete or *symbolic* dynamics and *continuous* dynamics. The main problem in the definition and representation of a hybrid system is precisely to specify the *interaction* between this symbolic and continuous dynamics.

A key issue in the formulation of hybrid systems is the often required *modularity* of the hybrid system description. Indeed, because we are inherently dealing with the modeling of *complex* systems, it is very important to model a complex hybrid system as the *interconnection* of simpler (hybrid) subsystems. This implies that the hybrid models that we are going to discuss are preferably of a form that admits easy interconnection and composition. Besides this notion of *compositionality* other important (related) notions are those of "reusability" and "hierarchy". These terms arise in the context of *"object-oriented modeling"*.

1.2 Towards a definition of hybrid systems

In our opinion, from a *conceptual* point of view the most basic definition of a hybrid system is to immediately specify its *behavior*, that is, the set of all possible trajectories of the continuous and discrete variables associated with the system. On the other hand, such a behavioral definition tends to be very general and far from an operational specification of hybrid systems. Instead we shall start with a reasonably generally accepted "working definition" of hybrid systems, which already has proved its usefulness. This definition, called the *hybrid automaton* model, will already provide the framework and terminology to discuss a range of typical features of hybrid systems. At the end of this chapter we are then prepared to return to the issues that enter into a behavioral definition of hybrid systems, and to discuss an alternative way of modeling hybrid systems by means of equations.

1.2.1 Continuous and symbolic dynamics

In order to motivate the hybrid automaton definition, we recall the "paradigms" of continuous and symbolic dynamics; namely, state space models described by differential equations for continuous dynamics, and finite automata for symbolic dynamics. Indeed, the definition of a hybrid automaton basically *combines* these two paradigms. Note that both in the continuous domain and in the discrete domain one can think of more general settings; for instance partial differential equations and stochastic differential equations on the continuous side, pushdown automata and Turing machines on the discrete side. The framework we shall discuss however is already suitable for the modeling of many applications, as will be shown in the next chapter.

Definition 1.2.1 (Continuous-time state-space models).
A continuous-time state-space system is described by a set of *state* variables x taking values in $\mathbb{R}^n$ (or, more generally, in an n-dimensional state space manifold X), and a set of *external* variables w taking values in $\mathbb{R}^q$, related by

a mixed set of differential and algebraic equations of the form

$$F(x, \dot{x}, w) = 0. \tag{1.1}$$

Here $\dot{x}$ denotes the derivative of x with respect to time. Solutions of (1.1) are all (sufficiently smooth) time functions $x(t)$ and $w(t)$ satisfying

$$F(x(t), \dot{x}(t), w(t)) = 0$$

for (almost) all times $t \in \mathbb{R}$ (the continuous-time axis).

Of course, the above definition encompasses the more common definition of a continuous-time input-state-output system

$$\begin{aligned} \dot{x} &= f(x, u) \\ y &= h(x, u) \end{aligned} \tag{1.2}$$

where we have split the vector of external variables w into a subvector u taking values in $\mathbb{R}^m$ and a subvector y taking values in $\mathbb{R}^p$ (with $m + p = q$), called respectively the vector of *input* variables and the vector of *output* variables. The only algebraic equations in (1.2) are those relating the output variables y to x and u, while generally in (1.1) there are additional algebraic constraints on the state space variables x.

One of the main advantages of general continuous-time state space systems (1.1) over continuous-time input-state-output systems (1.2) is the fact that the first class is closed under interconnection, while the second class in general is not. In fact, modeling approaches that are based on modularity (viewing the system as the interconnection of smaller subsystems) almost invariably lead to a mixed set of differential and algebraic equations. Of course, in a number of cases it may be relatively easy to eliminate the algebraic equations in the state space variables, in which case (if we can also easily split w into u and y) we can convert (1.1) into (1.2).

We note that Definition 1.2.1 does not yet completely specify the continuous-time system, since (on purpose) we have been rather vague about the precise *solution concept* of the differential-algebraic equations (1.1). For example, a reasonable choice (but not the only possible one!) is to require $w(t)$ to be piecewise continuous (allowing for discontinuities in the "inputs") and $x(t)$ to be continuous and piecewise differentiable, with (1.1) being satisfied for almost all t (except for the points of discontinuity of $w(t)$ and nondifferentiability of $x(t)$).

Next we give the standard definition of a finite automaton (or *finite state machine*, or *labeled transition system*).

Definition 1.2.2 (Finite automaton). A finite automaton is described by a triple (L, A, E). Here L is a finite set called the *state space*, A is a finite set called the *alphabet* whose elements are called *symbols*. E is the *transition rule*; it is a subset of $L \times A \times L$ and its elements are called *edges* (or transitions, or events). A sequence $(l_0, a_0, l_1, a_1, \ldots, l_{n-1}, a_{n-1}, l_n)$ with $(l_i, a_i, l_{i+1}) \in E$ for $i = 1, 2, \ldots, n - 1$ is called a trajectory or *path*.

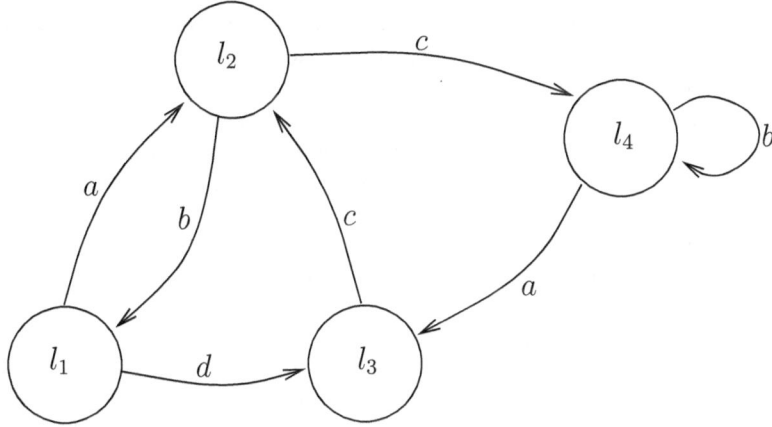

Figure 1.1: Finite automaton

The usual way of depicting an automaton is by a *graph* with vertices given by the elements of L, and edges given by the elements of E, see Figure 1.1. Then A can be seen as a set of *labels* labeling the edges. Sometimes these are called *synchronization* labels, since interconnection with other automata takes place via these (shared) symbols. One can also specialize Definition 1.2.2 to *input-output* automata by associating with every edge *two* symbols, namely an *input* symbol i and an *output* symbol o, and by requiring that for every input symbol there is only one edge originating from the given state with this input symbol. (Sometimes such automata are called *deterministic* input-output automata.) Deterministic input-output automata can be represented by equations of the following form:

$$
\begin{aligned}
l^{\sharp} &= \nu(l, i) \\
o &= \eta(l, i)
\end{aligned}
\tag{1.3}
$$

where $l^{\sharp}$ denotes the new value of the discrete state *after* the event takes place, resulting from the old discrete state value l and the input i.

Often the definition of a finite automaton also includes the explicit specification of a subset $I \subset L$ of *initial states* and a subset $F \subset L$ of *final states*. A path $(l_0, a_0, l_1, a_1, \ldots, l_{n-1}, a_{n-1}, l_n)$ is then called a *successful path* if in addition $l_0 \in I$ and $l_n \in F$.

In contrast with the continuous-time systems defined in Definition 1.2.1 the solution concept (or *semantics*) of a finite automaton (with or without initial and final states) is completely specified: the behavior of the finite automaton consists of all (successful) paths. In theoretical computer science parlance the definition of a finite automaton is said to entail an "operational semantics", completely specifying the formal language generated by the finite automaton.

Note that the definition of a finite automaton is conceptually not very different from the definition of a continuous-time state space system. Indeed we may relate the state space L with the state space X, the symbol alphabet A with the space W (where the external variables take their values), and the transition rule E with the set of differential-algebraic equations given by (1.1). Furthermore the paths of the finite automaton correspond to the *solutions* of the set of differential-algebraic equations. The analogy between continuous-time input-state-output systems (1.2) and input-output automata (1.3) is obvious, with the differentiation operator $\frac{d}{dt}$ replaced by the "next state" operator $\sharp$.

A (minor) difference is that in finite automata one usually considers (as in Definition 1.2.2) paths of *finite* length, while for continuous-time state space systems the emphasis is on solutions over the whole time axis $\mathbb{R}$. This could be remedied by adding to the finite automaton a *source state* and a *sink state* and a blank label, and by considering solutions defined over the whole time axis $\mathbb{Z}$ which "start" at minus infinity in the source state and "end" at plus infinity in the sink state, while producing the blank symbol when remaining in the source or sink state. Also the set I of initial states and the set F of final states in some definitions of a finite automaton do not have a direct analogon in the definition of a continuous-time state space system. In some sense, however, they could be viewed as a formulation of *performance* specifications of the automaton.

Summarizing, the basic differences between Definition 1.2.1 and Definition 1.2.2 are the following.

- The spaces L and A are *finite sets* instead of *continuous* spaces such as X and W. (In some extensions one allows for *countable* sets L and A.)

- The time axis in Definition 1.2.1 is $\mathbb{R}$, while the time axis in Definition 1.2.2 is $\mathbb{Z}$. Here, to be precise, $\mathbb{Z}$ is understood *without* any structure of addition (only sequential ordering).

- In the finite automaton model the set of possible transitions (events) is specified explicitly, while the evolution of a continuous-time state-space system is only *implicitly* given by the set of differential and algebraic equations (one still needs to *solve* these equations).

1.2.2   Hybrid automaton

Combining Definitions 1.2.1 and 1.2.2 leads to the following type of definition of a hybrid system.

Definition 1.2.3 (Hybrid automaton). A *hybrid automaton* is described by a septuple $(L, X, A, W, E, Inv, Act)$ where the symbols have the following meanings.

- L is a finite set, called the set of *discrete states* or *locations*. They are the *vertices* of a graph.

- X is the continuous state space of the hybrid automaton in which the continuous state variables x take their values. For our purposes $X \subset \mathbb{R}^n$ or X is an n-dimensional manifold.

- A is a finite set of symbols which serve to label the edges.

- $W = \mathbb{R}^q$ is the continuous communication space in which the continuous external variables w take their values.

- E is a finite set of edges called transitions (or *events*). Every edge is defined by a five-tuple $(l, a, Guard_{ll'}, Jump_{ll'}, l')$, where $l, l' \in L$, $a \in A$, $Guard_{ll'}$ is a subset of X and $Jump_{ll'}$ is a relation defined by a subset of $X \times X$. The transition from the discrete state l to l' is *enabled* when the continuous state x is in $Guard_{ll'}$, while during the transition the continuous state x jumps to a value x' given by the relation $(x, x') \in Jump_{ll'}$.

- Inv is a mapping from the locations L to the set of subsets of X, that is $Inv(l) \subset X$ for all $l \in L$. Whenever the system is at location l, the continuous state x must satisfy $x \in Inv(l)$. The subset $Inv(l)$ for $l \in L$ is called the *location invariant* of location l.

- Act is a mapping that assigns to each location $l \in L$ a set of differential-algebraic equations F_l, relating the continuous state variables x with their time-derivatives $\dot{x}$ and the continuous external variables w:

$$F_l(x, \dot{x}, w) = 0. \tag{1.4}$$

The solutions of these differential-algebraic equations are called the *activities* of the location.

Clearly, the above definition draws strongly upon Definition 1.2.2, the discrete state space L now being called the space of locations. (Note that the set of edges E in Definition 1.2.3 also defines a subset of $L \times A \times L$.) In fact, Definition 1.2.3 extends Definition 1.2.2 by associating with every vertex (location) a continuous dynamics whose solutions are the activities, and by associating with every transition $l \to l'$ also a possible jump in the continuous state.

Note that the *state* of a hybrid automaton consists of a discrete part $l \in L$ and a continuous part in X. Furthermore, the *external variables* consist of a discrete part taking their values a in A and a continuous part w taking their values in $\mathbb{R}^q$. Also, the dynamics consists of discrete transitions (from one location to another), together with a continuous part evolving in the location invariant.

It should be remarked that the above definition of a hybrid automaton has the same ambiguity as the definition of a continuous-time state-space system, since it still has to be complemented by a precise specification of the solutions (activities) of the differential-algebraic equations associated with every location. In fact, in the original definitions of a hybrid automaton (see e.g. [1]) the activities of every location are assumed to be explicitly given rather than generated implicitly as the solutions to the differential-algebraic equations. On the other hand, somebody acquainted with differential equations would not find it convenient in general to have to specify continuous dynamics immediately by time functions from $\mathbb{R}^+$ to X. Indeed, continuous time dynamics is almost always described by sets of differential or differential-algebraic equations, and only in exceptional cases (such as linear dynamical systems) one can obtain explicit solutions. The description of a hybrid automaton is illustrated in Figure 1.2.

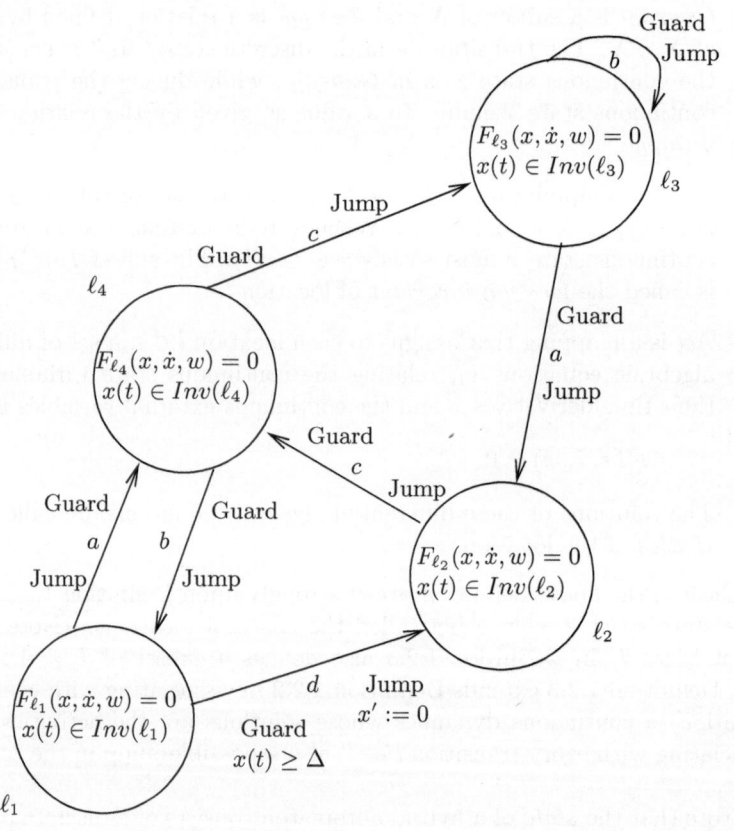

Figure 1.2: Hybrid automaton

A reasonable definition of the *trajectories* (or solutions, or in computer science terminology, the *runs* or *executions*) of a hybrid automaton can be formulated as follows. A continuous trajectory (l, δ, x, w) associated with a location l consists of a nonnegative time δ (the *duration* of the continuous trajectory), a piecewise continuous function $w : [0, \delta] \to W$, and a continuous and piecewise differentiable function $x : [0, \delta] \to X$ such that

- $x(t) \in Inv(l)$ for all $t \in (0, \delta)$,

- $F_l(x(t), \dot{x}(t), w(t)) = 0$ for all $t \in (0, \delta)$ except for points of discontinuity of w.

A *trajectory* of the hybrid automaton is an (infinite) sequence of continuous trajectories

$$(l_0, \delta_0, x_0, w_0) \overset{a_0}{\to} (l_1, \delta_1, x_1, w_1) \overset{a_1}{\to} (l_2, \delta_2, x_2, w_2) \overset{a_2}{\to} \ldots$$

such that at the *event times*

$$t_0 = \delta_0, \quad t_1 = \delta_0 + \delta_1, \quad t_2 = \delta_0 + \delta_1 + \delta_2, \ldots$$

the following inclusions hold for the discrete transitions:

$$\begin{aligned} x_j(t_j) &\in Guard_{l_j l_{j+1}} \\ (x_j(t_j), x_{j+1}(t_j)) &\in Jump_{l_j l_{j+1}} \end{aligned} \qquad \text{for all } j = 0, 1, 2, \ldots.$$

Furthermore, to the j-th arrow $\to$ in the above sequence (with j starting at 0) one associates a symbol (label) a_j, representing the value of the discrete "signal" at the j-th discrete transition.

1.2.3 Features of hybrid dynamics

Note that the trajectories of a hybrid automaton exhibit the following features. Starting at a given location the continuous part of the state evolves according to the continuous dynamics associated with this location, provided it remains in the location invariant. Then, at some time instant in $\mathbb{R}$, called an *event time*, an event occurs and the discrete part of the state (the location) *switches* to another location. This is an *instantaneous* transition which is *guarded*, that is, a necessary condition for this transition to take place is that the guard of this transition is satisfied. Moreover in general this transition will also involve a *jump* in the continuous part of the state. Then, after the instantaneous transition has taken place, the continuous part of the state, starting from this new continuous state, will in principle evolve according to the continuous dynamics of the new location. Thus there are two phenomena associated with every event, a *switch* and a *jump*, describing the instantaneous transition of, respectively, the discrete and the continuous part of the state at such an event time.

A basic issue in the specification of a hybrid system is the *specification of the events and event times*. First, the events may be *externally induced* via the labels (symbols) $a \in A$; this leads to *controlled* switchings and jumps. Secondly, the events may be *internally induced*; this leads to what is called *autonomous* switchings and jumps. The occurrence of internally induced events is determined by the guards and the location invariants. Whenever violation of the location invariants becomes imminent, the hybrid automation *has to* switch to a new location, with possibly a reset of the continuous state. At such an event time the guards will determine to *which* locations the transition is possible. (There may be more than one; furthermore, it may be possible to switch to the *same* location.)

If violation of the location invariants is not imminent then *still* discrete transitions may take place if the corresponding guards are satisfied. That is, if at a certain time instant the guard of a discrete transition is satisfied then this may cause an event to take place. As a result one may obtain a large class of trajectories of the hybrid automaton, and a tighter specification of the behavior of the hybrid automaton will critically depend on a more restrictive definition of the guards. Intuitively the location invariants provide *enforcing conditions* whereas the guards provide *enabling conditions*.

Many issues naturally come up in connection with the analysis of the trajectories of a hybrid automaton. We list a number of these.

- It could happen that, after some time t, the system ends up in a state $(l, x(t))$ from which there is no continuation, that is, there is no possible continuous trajectory from $x(t)$ and no possible transition to another location. In the computer science literature this is usually called *"deadlock"*. Usually this will be considered an undesirable phenomenon, because it means that the system is "stuck".

- The set of durations δ_i may get smaller and smaller for increasing i even to such an extent that $\sum_{i=0}^{i=\infty} \delta_i$ is finite, say τ. This means that τ is an accumulation point of event times. In the computer science literature this is called *Zeno* behavior (referring to Zeno's paradox of Achilles and the turtle). This need not always be an undesirable behavior of the hybrid system. In fact, as long as the continuous and discrete parts of the state will converge to a unique value at the accumulation point τ, then we can re-initialize the hybrid system at τ using these limits, and let the system run as before starting from the initial time τ. In fact, in Subsection 2.2.3 (the bouncing ball) we will encounter a situation like this.

- In principle the durations of the continuous trajectories are allowed to be zero; in this way one may cover the occurrence of *multiple events*. In this case the underlying time axis of the hybrid trajectory has a structure which is more complicated than $\mathbb{R}$ containing a set of event times: a certain time instant $t \in \mathbb{R}$ may correspond to a *sequence* of sequentially ordered transitions, all happening at the same time instant t which is then called a multiple event time. We refer e.g. to Subsection 2.2.6 where we also provide an example of an event with multiplicity ∞.

- It may happen that the hybrid system gets stuck at such a multiple event time by switching indefinitely between different locations (and not proceeding in time). This is sometimes called *livelock*. Such a situation occurs if a location invariant tends to be violated and a guarded transition takes place to a new location in such a way that the new continuous state does not satisfy the location invariants of the new location, while the guard for the transition to the old location is satisfied. In some cases this problem can be resolved by creating a *new* location, with a continuous dynamics that "averages" the continuous dynamics of the locations between which the infinite switching occurs. Filippov's notions of solutions of discontinuous vector fields can be interpreted in this way, cf. Subsection 2.2.7 and Chapter 3.

- In general the set of trajectories (runs) of the hybrid automaton may be very large, especially if the guards are not very strict. For certain purposes (e.g. verification) this may not be a problem, but in other cases one may wish the hybrid system to be *deterministic* or *well-posed*, in the sense of having *unique* solutions for given discrete and continuous "inputs" (assuming that we have split the vector w of continuous external variables into a vector of continuous inputs and outputs, and that every label a actually consists of an input label and an output label). Especially for *simulation* purposes this may be very desirable, and in this context one would often dismiss a hybrid model as inappropriate if it does not have this well-posedness property. On the other hand, nondeterministic discrete-event systems are very common in computer science. A simple example of a hybrid system not having unique solutions, at least from a certain point of view, is provided by the integrator dynamics

$$\dot{y} = u$$

in conjunction with the relay element

$$u = +1, \qquad y > 0$$
$$u = -1, \qquad y < 0$$
$$-1 \leq u \leq 1, \quad y = 0.$$

The resulting hybrid system (without external inputs) has three locations corresponding to the three segments of the ideal relay element. It is directly seen that for initial condition $y(0) = 0$ the system can evolve in either of these three locations, yielding the three solutions (i) $y(t) = t$, $u(t) = 1$, (ii) $y(t) = -t$, $u(t) = -1$, and (iii) $y(t) = 0$, $u(t) = 0$. Hence, if we specify the initial condition of this hybrid system only by its *continuous* state, which is not unreasonable in physical systems, then there are three solutions starting from the zero initial condition. More involved examples of the same type will be given in Chapter 4. A "physical" example of the same type exhibiting non-uniqueness of solutions is the classical example, due to Painlevé, of a stick that slides with one end on a table and that is subject to Coulomb friction acting at the contact point; see e.g. [31, p. 154].

- The continuous dynamics associated to a location may have "finite escape time", in the sense that the solution of the differential equations, or of the differential-algebraic equations, for some initial state goes to "infinity" (or to the boundary of the state space X) in *finite time*. This is a well-known phenomenon in nonlinear differential equations. A simple example is provided by the differential equation

$$\dot{x}(t) = 1 + x^2(t), \quad x(0) = 0,$$

a solution of which is $x(t) = \tan t$. (The framework of hybrid systems allows one, in principle, to model this as an event, by letting the "explosion time" ($\frac{\pi}{2}$ in the above example) be an event time, where the system may switch to a new location and jump to a new continuous state.)

- The solution concept of the continuous-time dynamics associated to a location may itself be problematic, especially because of the possible presence of algebraic constraints. In particular, in some situations one may want to associate *jump behavior* with these continuous-time dynamics (see e.g. Subsection 2.2.15). Within the hybrid framework this can be incorporated as internally induced events, where the system switches to the same location but is subject to a reset of the continuous state.

Remark 1.2.4. Of course, the definition of a hybrid automaton can still be generalized in a number of directions. A particularly interesting extension is to consider *stochastic* hybrid systems, such as the systems described by *piecewise-deterministic Markov processes*, see e.g. [40]. In this notion the event times are determined by the system reaching certain boundaries in the continuous state space (similar to the notion of location invariants), and/or by an underlying *probability distribution*. Furthermore, also the resulting discrete transitions together with their jump relations are assumed to be governed by a probability distribution.

1.2.4 Generalized hybrid automaton

In Definition 1.2.3 of a hybrid automaton there is still an apparent asymmetry between the continuous and the symbolic (discrete) part of the dynamics. Furthermore, the location invariants and the guards play strongly related roles in the specification of the discrete transitions. The following generalization of Definition 1.2.3 takes the location invariants and the set of edges E together, and symmetrizes the definition of a hybrid automaton. (The input-output version of this definition is due to [98].)

Definition 1.2.5 (Generalized hybrid automaton). A *generalized hybrid automaton* is described by a sixtuple (L, X, A, W, R, Act) where L, X, A, W and Act are as in Definition 1.2.3, and R is a subset of $(L \times X) \times (A \times W) \times (L \times X)$.

A continuous trajectory (l, a, δ, x, w) associated with a location l and a discrete external symbol a consists of a nonnegative time δ (the *duration* of the continuous trajectory), a piecewise continuous function $w : [0, \delta] \to W$, and a continuous and piecewise differentiable function $x : [0, \delta] \to X$ such that

- $(l, x(t), a, w(t), l, x(t)) \in R$ for all $t \in (0, \delta)$,

- $F_l(x(t), \dot{x}(t), w(t)) = 0$ for almost all $t \in (0, \delta)$ (exceptions include the points of discontinuity of w).

A *trajectory* of the generalized hybrid automaton is an (infinite) sequence

$$(l_0, a_0, \delta_0, x_0, w_0) \to (l_1, a_1, \delta_1, x_1, w_1) \to (l_2, a_2, \delta_2, x_2, w_2) \to \ldots$$

such that at the event times

$$t_0 = \delta_0, t_1 = \delta_0 + \delta_1, t_2 = \delta_0 + \delta_1 + \delta_2, \ldots$$

the following relations hold:

$$(l_j, x_j(t_j), a_j, w_j(t_j), l_{j+1}, x_{j+1}(t_j)) \in R, \text{ for all } j = 0, 1, 2, \ldots$$

The subset R can be related to the notions of the location invariants, guards, and jumps of Definition 1.2.3 in the following way. To each location l we associate the generalized location invariant

$$Inv(l) = \{(x, a, w) \in X \times A \times W \mid (l, x, a, w, l, x) \in R\}.$$

(Note that the symbols x and w denote here elements of X and W respectively rather than *variables* taking their values in these spaces.) Furthermore, given two locations l, l' we obtain the following generalized guard for the transition from l to l':

$$Guard_{ll'} = \{(x, a, w) \in X \times A \times W \mid \exists x' \in X, (l, x, a, w, l', x') \in R\}$$

with the interpretation that the transition from l to l' can take place if and only if $(x, a, w) \in Guard_{ll'}$. Finally, the associated generalized jump relation is given by

$$Jump_{ll'} = \{(x, x', a, w) \in X \times X \times A \times W \mid (l, x, a, w, l', x') \in R\}.$$

The resulting location invariants, guards as well as jump relations are of a more general type than was used in Definition 1.2.3, since they allow incorporation of constraints relating to the continuous and discrete external variables. (In Subsection 2.2.13 of Chapter 2 we shall see that this is actually an appropriate generalization.) Conversely, it can be readily seen that any set E of edges and location invariants as in Definition 1.2.3 can be recovered from a suitably chosen set R as in Definition 1.2.5. Therefore, Definition 1.2.5 does indeed generalize Definition 1.2.3; that is, any hybrid automaton in the sense of Definition 1.2.3 is also a hybrid automaton in the sense of Definition 1.2.5.

A further widening of the definition of a hybrid automaton would be to allow different continuous state spaces associated with different locations. (This is now to some extent captured in the location invariants or the subset R.) In fact, some of the examples that we will present in Chapter 2 do motivate such a generalization.

The definition of a (generalized) hybrid automaton admits *compositionality* in the following sense (cf. [1]). For simplicity we shall only give the definitions for the generalized hybrid automaton model (Definition 1.2.5). Consider two generalized hybrid automata $\Sigma_i = (L_i, X_i, A_i, W_i, R_i, Act_i)$, $i = 1, 2$, and suppose that

$$W_1 = W_2, \quad A_1 = A_2$$

(This is the case of *shared* external variables.) We define the *synchronous parallel composition* or *interconnection* $\Sigma = \Sigma_1 \parallel \Sigma_2$ of the two generalized hybrid automata $\Sigma_i, i = 1, 2$, as the generalized hybrid automaton (L, X, A, W, R, Act) given by

$$
\begin{aligned}
L &= L_1 \times L_2 \\
X &= X_1 \times X_2 \\
A &= A_1 = A_2 \\
W &= W_1 = W_2 \\
R &= \{((l_1, l_2), (x_1, x_2), a, w, (l_1', l_2'), (x_1', x_2')) \in (L \times X) \\
&\qquad \times (A \times W) \times (L \times X) \mid (l_i, x_i, a, w, l_i', x_i') \in R_i, i = 1, 2\} \\
Act &= Act_1 \times Act_2.
\end{aligned}
\tag{1.5}
$$

This corresponds to the *full* synchronous parallel composition or interconnection of the two generalized hybrid automata. We may also consider *partial* synchronous parallel compositions by synchronizing only a *part* of the discrete external variables a_1 and in a_2 in A_1, respectively A_2, and by considering more general relations between the continuous external variables w_1 and w_2. Furthermore, we may also define the *interleaving parallel composition* $\Sigma = \Sigma_1 \parallel\mid \Sigma_2$ by taking $A = A_1 \times A_2$ and $W = W_1 \times W_2$ in (1.5), and by defining $R = R_1 \times R_2$.

1.2.5   Hybrid time evolutions and hybrid behavior

A conceptual problem in Definitions 1.2.3 and 1.2.5 is the formalization of the notion of the time evolution corresponding to a hybrid trajectory, and in particular the embedding of the event times in the continuous time axis $\mathbb{R}$. In Def. 1.2.5 time is broken up into a sequence of closed intervals, which may reduce to single points. Effectively a counting label is added to time points which indicates to which element of the sequence the point belongs; this is necessary to make it possible that state variables have different values at the "same" event time. A similar labeling procedure has been used for instance in [162], where the authors speak of a "time space". These notions of time are not able to cover situations in which the event times of a trajectory have an accumulation point but still the trajectory does progress in time *after* this accumulation point. Examples of such a situation are provided in Subsections 2.2.3 and 2.2.6. Therefore we propose here a somewhat more general notion.

Let $\mathbb{R}$ be the continuous time axis. The *time evolution* corresponding to a hybrid trajectory will be specified by a set $\mathcal{E}$ of *time events*. A time event in $\mathcal{E}$ consists of an *event time* $t \in \mathbb{R}$ together with a *multiplicity* $m(t)$, which is an element of $\mathbb{N} \cup \{\infty\}$, where $\mathbb{N}$ is the set of natural numbers $1, 2, 3, \ldots$. The time event will be denoted by the *sequence*

$$(t^{0\sharp}, t^{1\sharp}, t^{2\sharp}, \ldots, t^{m(t)\sharp})$$

specifying the sequentially ordered "discrete transition times" at the *same* continuous time instant $t \in \mathbb{R}$ (the event time). For simplicity of notation we will sometimes write $t^{\sharp}, t^{\sharp\sharp}, t^{\sharp\sharp\sharp}, \ldots$ for $t^{1\sharp}, t^{2\sharp}, t^{3\sharp}, \ldots$.

A time event with multiplicity equal to 1 is just given by a pair

$$(t^{0\sharp}, t^{\sharp})$$

with the interpretation of denoting the time instants "just before" and "just after" the event has taken place (a more formal description will be given in a moment). If the multiplicity of the time event is larger than 1 (a *multiple time event*) then there are some "intermediate time instants" ("all at the same event time t") ordering the sequence of discrete transitions taking place at t.

The "embedding" of the discrete dynamics into the continuous dynamics will be performed by the compatibility conditions

$$\begin{aligned} x(t^{+}) &= x(t^{m(t)\sharp}) \\ x(t^{-}) &= x(t^{0\sharp}), \end{aligned}$$

where, in a usual notation,

$$\begin{aligned} x(t^{+}) &:= \lim_{\tau \downarrow t} x(\tau) \\ x(t^{-}) &:= \lim_{\tau \uparrow t} x(\tau). \end{aligned}$$

Note that events are allowed to have multiplicity equal to ∞, in which case there are an infinite number of discrete transitions taking place at the same continuous time instant t. It seems natural to require that in such a situation the sequence of locations $l_1, l_2, l_3, \ldots$ at this time t converges to a single location, and that $\lim_{k \to \infty} x(t^{k\sharp})$ exists (or, perhaps, has only a finite number of accumulation points). An example of such a situation will be provided in Subsection 2.2.6.

The set of event times corresponding to a set of time events $\mathcal{E}$ will be denoted by $\mathcal{E}_{\mathcal{T}} \subset \mathbb{R}$. Hence the set $\mathcal{E}$ of time events can be written as

$$\mathcal{E} = \cup_{t \in \mathcal{E}_{\mathcal{T}}} \{(t^{0\sharp}, t^{1\sharp}, \ldots, t^{m(t)\sharp})\}.$$

For the moment we will allow $\mathcal{E}_\mathcal{T}$ to be an arbitrary subset of $\mathbb{R}$, but later on (when dealing with the solution concept of a hybrid system) we will put restrictions on $\mathcal{E}_\mathcal{T}$.

The *time evolution* $\tau_\mathcal{E}$ corresponding to a time event set $\mathcal{E}$ is defined to be the set

$$\tau_\mathcal{E} = (\mathbb{R} \setminus \mathcal{E}_\mathcal{T}) \cup \mathcal{E}.$$

The union of all time evolutions $\tau_\mathcal{E}$ for all time event sets $\mathcal{E}$ will be denoted by $\mathcal{T}$. The notion of a time evolution is illustrated in Figure 1.3.

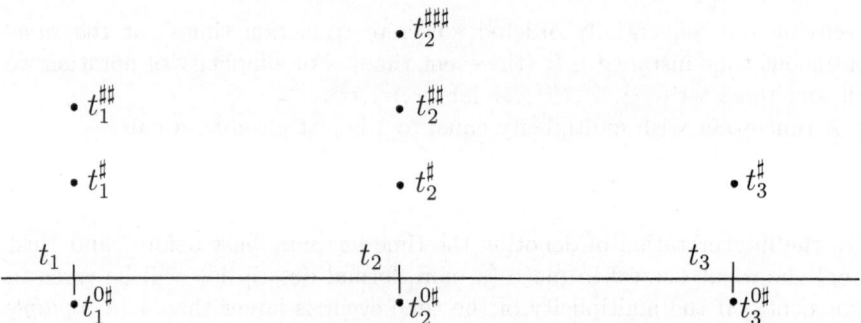

Figure 1.3: Time evolution

We now formalize the notion of a trajectory of a generalized hybrid automaton using the concept of time evolutions. For clarity of notation, we will distinguish the elements $l \in L$ and $a \in A$ from the *variables* P, respectively S, taking their *values* in L, respectively A.

A *trajectory* of the generalized hybrid automaton is then defined by a time event set $\mathcal{E}$, together with the corresponding time evolution $\tau_\mathcal{E}$, and functions $P : \tau_\mathcal{E} \to L$, $x : \tau_\mathcal{E} \to X$, $S : \mathcal{E} \to A$, $w : \tau_\mathcal{E} \to W$, such that

- *discrete dynamics*: for every time event $(t^{0\sharp}, t^{1\sharp}, t^{2\sharp}, \ldots, t^{m(t)\sharp})$

$$(P(t^{k\sharp}), x(t^{k\sharp}), S(t^{k\sharp}), w(t^{k\sharp}), P(t^{(k+1)\sharp}), x(t^{(k+1)\sharp})) \in R,$$
$$k = 0, 1, \ldots, m(t) - 1$$

Furthermore $P(\tilde{t}) = P(t^{m(t)\sharp})$, for all $\tilde{t} > t$ up to the next event time t', as well as $P(t'^{0\sharp}) = P(t^{m(t)\sharp})$, and $S(t'^{0\sharp}) = S(t^{m(t)\sharp})$.

(Here it is assumed that for all event times t the multiplicity $m(t)$ is either finite, or, in case $m(t) = \infty$, that $\lim_{k \to \infty} x(t^{k\sharp})$ and $\lim_{k \to \infty} P(t^{k\sharp})$ both exist, in which case we denote these limits by $x(t^{m(t)\sharp})$, respectively $P(t^{m(t)\sharp})$.)

- *continuous dynamics*: for all event times t_1, t_2 such that the interval (t_1, t_2) has zero intersection with $\mathcal{E}_T$ the function $w(t)$ is piecewise continuous and the function x is continuous and piecewise differentiable on (t_1, t_2) and satisfies

$$F_{l(t_1)}(x(t), \dot{x}(t), w(t)) = 0, \quad \text{for almost all} \quad t \in (t_1, t_2)$$
$$x(t_1^+) = x(t_1^{m(t_1)\sharp})$$
$$x(t_2^-) = x(t_2^{0\sharp})$$

whereas

$$(P(t_1^{m(t)\sharp}), x(t), S(t_1^{m(t)\sharp}), w(t), P(t_1^{m(t)\sharp}), x(t)) \in R,$$
$$\text{for all } t \in (t_1, t_2).$$

(Here, as before, $x(t_1^+) := \lim_{t\downarrow t_1} x(t)$ and $x(t_2^-) := \lim_{t\uparrow t_2} x(t)$.)

The above formalization of time evolutions of trajectories of hybrid systems also enables us to state a *behavioral* definition of a hybrid system, as follows.

Definition 1.2.6 (Hybrid behavior). Let $W = \mathbb{R}^q$ be a continuous communication space and let A be a finite communication space. Define the *universum* $\mathcal{U}$ of all possible trajectories given by triples $(\tau_{\mathcal{E}}, a, w)$, with $\tau_{\mathcal{E}}$ the time evolution corresponding to some time event set $\mathcal{E}$, and functions

$$S : \mathcal{E} \to A$$
$$w : \tau_{\mathcal{E}} \to W$$

with the property that at all event times $t \in \mathcal{E}_T$ with finite multiplicity $m(t)$ the discrete variables satisfy $S(t^{m(t)\sharp}) = S((t')^{0\sharp})$ where t' is the subsequent event time. A hybrid system with continuous variables taking values in W and discrete variables taking values in A is now defined to be a *subset* $\mathcal{B}$ of $\mathcal{U}$, and is denoted by $(W, A, \mathcal{B})$.

For many purposes one will actually adapt the above definition by shrinking the universum $\mathcal{U}$. In particular one may want to restrict to functions w which have some regularity properties like continuity or (piecewise) differentiability. Furthermore, one may wish to restrict the admissible sets of event times $\mathcal{E}_T$. An easy choice is to restrict to sets $\mathcal{E}_T$ consisting only of *isolated* points having no accumulation points. This excludes however examples like the bouncing ball in the next chapter (Subsection 2.2.3). Allowing for accumulation points, on the other hand, creates some problems for the compositionality of hybrid behaviors (because the accumulation points are removed from the continuous time axis). We come back to these issues later on.

Remark 1.2.7. The *projection* of the hybrid behavior on the behavior of the discrete variables S is is given by the values taken by the discrete variables at the time events, that is, on the set of time instants $\ldots, t_1^{0\sharp}, t_1^{\sharp}, t_1^{\sharp\sharp}, \ldots, t_1^{m(t_1)\sharp}, t_2^{0\sharp}, t_2^{\sharp}, t_2^{\sharp\sharp}, \ldots, t_2^{m(t_2)\sharp}, t_3^{0\sharp}, t_3^{\sharp}, \ldots$, where $\ldots, t_1, t_2, t_3, \ldots$ are event times. Moreover, if $m(t_i)$ is finite, then $S(t_{i+1}^{0\sharp}) = S(t_i^{m(t_i)\sharp})$. This defines the behavior of a *discrete-event system*.

Note, however, that in general the set of "discrete time instants" $\ldots, t_1^{0\sharp}, t_1^{\sharp}, t_1^{\sharp\sharp}, \ldots, t_1^{m(t_1)\sharp}, t_2^{0\sharp}, t_2^{\sharp}, t_2^{\sharp\sharp}, \ldots, t_2^{m(t_2)\sharp}, t_3^{0\sharp}, t_3^{\sharp}, \ldots$ of this resulting discrete-event system may have a very complicated structure.

Analogously, the projection of the hybrid behavior on the behavior of the continuous variables defines a continuous-time behavior defined on the time axis $\mathbb{R}$ minus the event times and their accumulation points.

1.2.6   Event-flow formulas

In this section we provide an alternative framework for modeling hybrid systems, which is *equation-based* and which is therefore in some sense closer to the usual modeling frameworks for continuous systems than to the graph-related representations that are often used for discrete systems. Guards, invariants and discrete transitions are summarized in Definition 1.2.5 in one abstract set R. Such a formulation has the advantage of being general, but in practice the set R will usually be described by equations (taken in a general sense to include also inequalities). Adjoining these equations to the usual sets of differential and algebraic equations leads to a description in terms of what we shall call *event-flow formulas*. Such an equation-based framework may be an attractive way of modeling hybrid systems with a substantial continuous-time dynamics; see also the discussion in Subsection 1.2.8.

The style of description in the methodology of event-flow formulas is somewhat similar to the way in which "sentences" are described in model theory. We begin by listing the types of variables that may occur. A precise semantics will be given later, but we already indicate the intended meaning of the symbols in order to facilitate the exposition. We use the word "variable" below to denote symbols that will be subject to evaluation at time points t in the semantics to be given below. We consider expressions formed from variables of the following types:

- continuous state variables, denoted by $x_1, x_2, \ldots, x_n$

- discrete state variables, denoted by $P_1, P_2, \ldots, P_k$

- continuous communication variables, denoted by $w_1, w_2, \ldots, w_q$

- discrete communication variables, denoted by $S_1, S_2, \ldots, S_r$.

The vector $(x_1, \ldots, x_k)$ will be abbreviated by x, and a similar notational convention will be used for the other collections of variables. In the semantics, we will associate to the symbol x a vector of real-valued functions of time taking values in a space X. Likewise, the continuous communication variables w will take values in a space W. Each of the discrete state variables P_i will take values in a finite set L_i; the product $L := L_1 \times \cdots \times L_k$ is the set of "locations". Finally, the discrete communication variables S_i take values in an alphabet A_i, and the product $A_1 \times \cdots \times A_r$ will be denoted by A. As a simple mnemonic device, the discrete variables are denoted by capital letters P and S, while the continuous variables x and w are in lowercase. As noted before, the notation also serves to emphasize the nature of a "discrete variable" *taking values* in the discrete spaces L and A, rather than being an *element* of these discrete spaces.

The communication variables may be used to link several parts of a system description to each other. One may also consider "open" systems in which the behavior of the communication variables is not completely determined by the system description itself; in such cases the communication variables may be thought of as providing a link to the (unmodeled) outside world.

To the symbols introduced above we associate certain other symbols, namely:

- for each continuous state variable x_i, there is also a variable $\dot{x}_i$ (the *derivative* of x_i)

- for each continuous state variable x_i and each discrete state variable P_i there are variables $x_i^{\sharp}$ and $P_i^{\sharp}$ (the *next value*).

We consider expressions in all the above mentioned variables, which are Boolean combinations of what we call *flow clauses* and *event clauses*. For our purposes it seems enough to consider *flow clauses* which are either of the form

$$\phi(\dot{x}, x, w, P) \;=\; 0 \tag{1.6}$$

(equality type) or of the form

$$\phi(x, w, P) \;\geq\; 0 \tag{1.7}$$

(inequality type), where in both cases ϕ is a real-valued function defined on the appropriate domain. Furthermore, we consider *event clauses* which are of the form

$$\phi(x^{\sharp}, P^{\sharp}, x, P, S) \;\geq\; 0 \tag{1.8}$$

where again ϕ is a real-valued function defined on the appropriate domain. The following definition expresses the notion that at each time the systems that we consider are subject either to a flow or to an event.

Definition 1.2.8. An *event-flow formula*, or EFF, is a Boolean formula whose terms are clauses, and which can be written in the form $F \vee E$ where F is a Boolean combination of flow clauses and E is a Boolean combination of event clauses.

Remark 1.2.9. Comparing with the definition of a generalized hybrid automaton (Definition 1.2.5), we see that the flow clauses determine the differential-algebraic equations describing the activities together with "a part" of the subset R (namely, the part involved in the specification of the continuous dynamics), while the event clauses determine the part of R specifying the discrete dynamics.

Remark 1.2.10. We shall freely use alternative notations in cases when it is clear how these can be fitted in the above framework. For instance, we write down equalities in event conditions even when the above formulation gives only inequalities, since an equality can be constructed from two inequalities. Also, when for instance P is a discrete variable that may take the values **on** and **off**, we write "$P = $ **on**" rather than first defining a function ϕ from the two-element set $\{$**on**, **off**$\}$ to $\mathbb{R}$ that takes the value 0 on **on** and 1 on **off**, and then writing "$\phi(P) = 0$".

Remark 1.2.11. For events with multiplicity 1 it is often notationally easier to use the variables x_i^- (the *left-hand limit*) and x_i^+ (the *right-hand limit*), and analogously P^- and P^+, to express the values *before* and *after* the event has taken place. This means that we also use instead of (1.8) event clauses of the form

$$\phi(x^+, P^+, x^-, P^-, S) \geq 0. \tag{1.9}$$

Remark 1.2.12. To further lighten the notation we shall often use a comma to represent conjunction, which is a standard convention actually. We use a vertical bar to denote disjunction between successive lines, so that

$$\left|\begin{array}{l} \text{clause}_1 \\ \text{clause}_2 \end{array}\right.$$

is read as

   clause$_1$ $\vee$ clause$_2$.

We shall also use indexed disjunctions, writing "$|_{i \in \{1,\dots,k\}}$clause$_i$" rather than "clause$_1$ $\vee$ $\cdots$ $\vee$ clause$_k$".

For the description of complex systems, it is essential that a composition operation is available which makes it possible to combine subsystems into larger systems. For flow conditions, such a composition may be based simply on conjunction (logical "and"), to express the intuition that the time we are dealing with here is physical time so that it should be common to all subsystems. Things are different however in the case of event conditions. If an event occurs in one subsystem, there are not necessarily events in all other subsystems; or it may happen that at the same physical time instant there are unrelated events, perhaps of different multiplicities, in several subsystems. It is therefore useful to introduce the notion of an "empty event", which is defined as follows. All of the alphabets A_i are extended with an element `blank` that is different from the existing elements; the interpretation of this value is "no signal". The *empty event* is given by the clause

$$x^+ = x^-, \quad P^+ = P^-, \quad w^+ = w^-, \quad S = \text{blank}. \tag{1.10}$$

We now consider the composition of a number of EFFs. So we start with a number of event-flow formulas $F_i \vee E_i$ ($i = 1, \ldots, \ell$), which are thought of as descriptions of subsystems. All of these subsystems have their own collections of symbols which however need not be disjoint. The disjunction of E_i with the empty event will be denoted by E_i'. The *parallel composition* (or just composition) of the subsystems given by $F_i \vee E_i$ is now defined as the EFF

$$(F_1 \vee E_1) \,||\, (F_2 \vee E_2) \,||\, \cdots \,||\, (F_\ell \vee E_\ell) \; :=$$
$$= (F_1 \wedge F_2 \wedge \cdots \wedge F_\ell) \vee (E_1' \wedge E_2' \wedge \cdots \wedge E_\ell'). \tag{1.11}$$

In this description of composition, communication between subsystems may take place by shared variables as well as by shared actions.

Event-flow formulas may be used for describing hybrid systems in a similar way as differential equations are used for describing smooth dynamical systems. As in the latter case, an exact interpretation of the equations requires the concept of a *solution*. In continuous systems, a choice has to be made here as to what function space will be used. In the hybrid system context, we still have this question but we also face a few more: in particular, to what extent will event times be allowed to accumulate, and may multiple events occur at the same time instant? Already in the case of smooth dynamical systems, there is no such thing as a "correct" answer to the question to what space the solutions of a given differential equation should belong; although some spaces are more popular than others, there is no unique choice that is good for all purposes and so in practice choices may vary depending on context. There is no reason to expect that the situation will be different for hybrid systems. We list a few choices that may turn out to be useful. In each case, we use a continuous state space X, a continuous communication space W, a discrete state space L, and a discrete communication space A. It is convenient to use some topological concepts.

Recall that a "time evolution" is a set of the form

$$\tau_\mathcal{E} \; = \; (\mathbb{R} \setminus \mathcal{E}_T) \cup \mathcal{E}$$

with

$$\mathcal{E} \; = \; \cup_{t \in \mathcal{E}_T} \{(t^{0\sharp}, t^{1\sharp}, \ldots, t^{m(t)\sharp})\}$$

where $\mathcal{E}_T$ is a subset of $\mathbb{R}$ and m is a function from $\mathcal{E}_T$ to $\mathbb{N} \cup \{\infty\}$. We say that the time evolution $\tau_\mathcal{E}$ and the set of time events $\mathcal{E}$ are *specified* by the pair $(\mathcal{E}_T, m)$. In order to ease the exposition we *restrict throughout* to sets $\mathcal{E}_T$ that are *closed and nowhere dense* subsets of $\mathbb{R}$. As before, we sometimes write $t^\sharp, t^{\sharp\sharp}, \ldots$ for $t^{1\sharp}, t^{2\sharp}, \ldots$. Furthermore, instead of $t^{0\sharp}$ and $t^{m(t)\sharp}$ we also sometimes write t^- and t^+ respectively. Sets of the form $(t - \delta, t) \cup \{t^-\}$ (with $t \in \mathcal{E}_T$ and $\delta > 0$) will be written as $(t - \delta, t^-]$, and likewise we will write $[t^+, t + \delta)$ instead of $(t, t + \delta) \cup \{t^+\}$. A time evolution $\tau_\mathcal{E}$ specified by a pair $(\mathcal{E}_T, m)$ is equipped with a *topology* generated by subsets of the following four forms: $(t_1, t_2) \subset \mathbb{R} \setminus \mathcal{E}_T$; $(t - \delta, t^-]$ with $t \in \mathcal{E}_T$ and $(t - \delta, t) \subset \mathbb{R} \setminus \mathcal{E}_T$; $[t^+, t + \delta)$ with $t \in \mathcal{E}_T$ and $(t, t + \delta) \subset \mathbb{R} \setminus \mathcal{E}_T$; $\{t^{j\sharp}\}$ with $t \in \mathcal{E}_T$ and $0 < j < m(t)$. (For the fundamental mathematical notion of "topology", see for instance [79].) The state spaces X and L and the communication spaces W and A are given their usual topologies; in the case of the discrete spaces this means the discrete topology, so that all points of these spaces are viewed as being isolated.

We now first define a space of trajectories that is about as large as one can have if one wants to be able to give a meaning to the expressions in an EFF.

Definition 1.2.13. The space $Z/\infty/C^1/L^1_{loc}$ consists of tuples $(\tau_\mathcal{E}, x, w, P, S)$, where

- $\tau_\mathcal{E}$ is a time evolution specified by a pair $(\mathcal{E}_T, m)$

- x is a continuous function from $\tau_\mathcal{E}$ to X which is continuously differentiable on $\mathbb{R} \setminus \mathcal{E}_T$

- w is a locally integrable function from $\mathbb{R} \setminus \mathcal{E}_T$ to W having lefthand and righthand limits at all points of $\mathcal{E}_T$

- P is a continuous function from $\tau_\mathcal{E}$ to L

- S is a function from $\mathcal{E}$ to A.

Remark 1.2.14. Since L is endowed with the discrete topology, continuity of P means that P is constant outside the event times.

Remark 1.2.15. The letter Z (for *Zeno*) refers to the fact that in the definition the set of event times $\mathcal{E}_T$ is allowed to have accumulation points; for instance $\mathcal{E}_T = \{\frac{1}{n} \mid n \in \mathbb{Z} \setminus \{0\}\} \cup \{0\}$, or even the Cantor set, could be sets of event times. Nevertheless, since $\mathcal{E}_T$ is assumed to be closed, these accumulation points are necessarily elements of $\mathcal{E}_T$. (If certain accumulation points are *not* in $\mathcal{E}_T$, e.g. if we would exclude the accumulation point $\{0\}$ from the above example of $\mathcal{E}_T$, then problems come up in defining the continuous dynamics starting at such an accumulation point.) The symbol ∞ indicates that event times can be of arbitrarily high or even infinite multiplicity. The symbols C^1 and L^1_{loc} indicate the degree of smoothness that is required for the trajectories of the continuous state variables and the continuous communication variables on the open set $\mathbb{R} \setminus \mathcal{E}_T$. The requirement concerning the existence of lefthand and righthand limits at event times (for the state variables this follows from the continuity requirement) is not made explicit in the notation since we will always impose such a condition.

Remark 1.2.16. Note that the definition allows the communication variables w to jump at event times, i.e. we do not necessarily have $w(t^-) = w(t^+)$ for $t \in \mathcal{E}_T$.

The space we have just introduced is very general; however, we shall often want to work with smaller and more manageable spaces. An example of such a space is the following.

Definition 1.2.17. The space $\mathrm{NZ}/1/C^{1/0}/C^0$ consists of tuples $(\tau_{\mathcal{E}}, x, w, P, S)$, where

- $\tau_{\mathcal{E}}$ is a time evolution specified by a pair $(\mathcal{E}_T, m)$ where $\mathcal{E}_T$ is a set of isolated points in $\mathbb{R}$, and $m(t) = 1$ for all $t \in \mathcal{E}_T$

- x is a continuous function from $\tau_{\mathcal{E}}$ to X which is continuously differentiable on $\mathbb{R} \setminus \mathcal{E}_T$ and which satisfies $x(t^+) = x(t^-)$ for all $t \in \mathcal{E}_T$

- w is a continuous function from $\mathbb{R} \setminus \mathcal{E}_T$ to W

- P is a continuous function from $\tau_{\mathcal{E}}$ to L

- S is a function from $\mathcal{E}$ to A.

Remark 1.2.18. The acronym NZ in the above notation stands for *non-Zeno*; it refers to the condition that the points of $\mathcal{E}$ are isolated, i.e. there are no accumulation points. The number 1 indicates that events have multiplicity one so there is no necessity to define intermediate states. The notation $C^{1/0}$ indicates that the continuous state variables are differentiable between events and continuous across events. Note that in this solution concept discontinuities in the external variables w (for instance in the input u) necessarily correspond to events.

We also define a third space, which is in some aspects more restricted and in other aspects more general than the previous one. The abbreviation RZ used below stands for "right-Zeno".

Definition 1.2.19. The space $\mathrm{RZ}/1/C^{1/0}$ consists of tuples $(\tau_{\mathcal{E}}, x, P)$, where

- $\tau_{\mathcal{E}}$ is a time evolution specified by a pair $(\mathcal{E}_T, m)$ where $\mathcal{E}_T$ is a set of *right isolated* points in $\mathbb{R}$, i.e. for every $t \in \mathcal{E}_T$ there is a $\delta > 0$ such that $(t, t + \delta) \cap \mathcal{E}_T = \emptyset$, and $m(t) = 1$ for all $t \in \mathcal{E}_T$

- x is a continuous function from $\tau_{\mathcal{E}}$ to X which is continuously differentiable on $\mathbb{R} \setminus \mathcal{E}_T$

- P is a continuous function from $\tau_{\mathcal{E}}$ to L.

Remark 1.2.20. This function class incorporates no communication variables, so as a space for system specification it is suitable only for "monolithic" systems (i.e. systems that are described as one closed whole, without use of subsystems and without communication to an outside world). The "bouncing ball" discussed in Subsection 2.2.3 is an example of a system that has solutions in the space $RZ/1/C^{1/0}$; the analogous space $NZ/1/C^{1/0}$ would not be suitable for this example.

Various other function spaces might be defined. Since any attempt at completeness would be futile, we do not list any further examples, but we shall feel free to use variants below.

At this point we can discuss the notion of solution for EFFs. The notion of solution is based on the evaluation of the elementary clauses in an EFF at specific time points. To be precise, a clause of the form $\phi(\dot{x}, x, w, P) = 0$ will be said to evaluate to TRUE for an element of $Z/\infty/C^1/L_{loc}^1$ at a time $t \in \mathbb{R} \setminus \mathcal{E}_T$ if

$$\phi(\dot{x}(t), x(t), w(t), P(t)) = 0.$$

Likewise, a clause $\phi(x, w, P) \geq 0$ evaluates to TRUE at $t \in \mathbb{R} \setminus \mathcal{E}_T$ if

$$\phi(x(t), w(t), P(t)) \geq 0.$$

An event clause $\phi(x^\sharp, P^\sharp, x, P, S) \geq 0$ evaluates to TRUE at a time $t^{j\sharp} \in \mathcal{E}$ if either $j < m(t)$ and

$$\phi(x(t^{(j+1)\sharp}), P(t^{(j+1)\sharp}), x(t^{j\sharp}), P(t^{j\sharp}), S(t^{j\sharp})) \geq 0,$$

or $j = m(t)$. Clearly, when clauses have well-defined truth values then any Boolean combination also has a well-defined truth value. After these preparations, the notion of solution can be defined as follows.

Definition 1.2.21. An element $(\tau_\mathcal{E}, x, w, P, S)$ of $Z/\infty/C^1/L_{loc}^1$ is said to be a *solution* of a given event-flow formula if the flow condition evaluates to TRUE for all $t \in \mathbb{R} \setminus \mathcal{E}_T$ and the event condition evaluates to TRUE for all $t^{j\sharp} \in \mathcal{E}$.

Similar definitions can be formulated for other time/trajectory spaces. The fact that in principle different definitions have to be given for different function spaces is already common in the theory of ordinary differential equations.

Remark 1.2.22. Sometimes a system description can be considerably simplified by using what might be called a *persistent-mode convention*, which enforces that mode changes will occur only when they are necessary to prevent violation of flow conditions. Typically the involved conditions are expressed as inequality constraints on the continuous state variables. In the mechanics literature the persistent-mode convention is sometimes known as "Kilmister's principle" [86], [31]. Formally, a PMC solution for a given event-flow formula is a solution having the property that, for each t_0, there is no solution of the same EFF that coincides with the given solution for $t < t_0$, that is defined for $t < t_1$ for some $t_1 > t_0$, and that has no event at t_0.

1.2.7 Simulation of hybrid systems

After specifying a complete model for a hybrid system, that is, a syntactically correct model together with a univocal semantics, it can be used for analysis, simulation and control.

For simulation purposes it is natural to require that the hybrid model under consideration has *unique* solutions for every initial state and every external "input" signal. Such hybrid systems have been called *"well-posed"*, and it is important to derive verifiable conditions which ensure well-posedness.

Once it has been established that a particular system is well-posed in the sense that trajectories are uniquely defined at least on some interval, the next question arises of actually *computing* the solutions. Although in principle any constructive existence proof for solutions might be used as a basis for calculation, the requirements imposed by numerical efficiency and theoretical rigor are quite different and so methods of computation may differ from methods of proof. This is certainly the case for smooth dynamical systems, where for instance a Picard iteration may be used for a proof of existence (which obtains the solution as a limit of a provably converging sequence of functions), whereas for simulation purposes one would typically use a stepping algorithm in which the difference between the value of the state vector at some time and its value at the next time step is approximated on the basis of a few computed function values. In the same way, the simulation of hybrid systems is not necessarily tied to the theory of well-posedness of such systems. Nevertheless the well-posedness theory does provide a starting point, and the combination of algorithms for mode selection and jump determination with standard methods for the simulation of smooth dynamics may already lead to workable simulation routines. Much remains to be worked out in this area however and in this subsection we will only give a very brief outline.

Basically there are three different approaches to the simulation of hybrid systems, which each may be worked out in many different ways. The three approaches may be briefly described as follows (cf. [114]).

1. *The smoothing method.* In this method, one tries to replace the hybrid model by a smooth model which is in some sense close to it. For instance, diodes in an electrical network may be described as ideal diodes (possibly plus some other elements), which will give rise to regime-switching dynamics, or as strongly nonlinear resistors, which gives rise to smooth dynamics. Similarly, in a mechanical system with unilateral constraints one might describe collisions as instantaneous, and then one must allow jumps in velocities; or one might describe them in terms of a compression and a decompression phase, and in that case jumps in velocities may be avoided. In a similar way, unilateral constraints in an optimization problem may be replaced by penalty functions. Taking an extreme point of view, one might argue that even switches in computer-controlled processes may be described by smooth dynamical systems; indeed, the transistors inside the computer that physically carry out the switching can be described for instance in terms of the Ebers-Moll differential

equations (see e. g. [46, p. 724]). Reasoning in this way, a point could be made that it is almost always possible to provide a smooth model that is "closer to reality" than a competing hybrid model.

Nevertheless, one can easily come up with examples that would be relatively awkward to describe on the basis of smoothing. If one looks at a bouncing ping-pong ball on a flat table, then the nonsmooth model comes to mind immediately since the time during which the ball is in contact with the table is very short in comparison with the flight phases. In a smooth model one would be forced to spend considerable effort in specifying various parameters which in the nonsmooth model are all captured by one dominant and easily observable parameter, namely the restitution coefficient. Indeed, the strength of a hybrid model as opposed to a smooth model is usually the simplicity of the first.

Obviously the discussion about whether to use smooth or nonsmooth models should be related to the actual purpose of modeling. If a model for a bouncing ball is formulated with the aim of predicting the number of bounces that will occur, then the simple nonsmooth model with a constant coefficient of restitution is not of much use, since it predicts an infinite number of bounces — an answer which is incorrect and, even worse, not informative. At a general level of discussion one cannot make an argument for one or the other. But one can say that at least for certain problems nonsmooth models are more convenient than smooth models. Since in this book we are interested in nonsmooth modeling, we shall further limit the discussion to that case. We shall not even spend much attention to the interesting questions around the convergence of solutions of smooth models to solutions of nonsmooth models.

2. *The event-tracking method.* The most common way of generating trajectories of hybrid systems seems to be the one that is based on the following sequence:

(i) simulation of the smooth dynamics within a given mode (discrete state);

(ii) event detection;

(iii) determination of a new discrete state (new mode);

(iv) determination of a new continuous state (re-initialization).

The idea is to simulate the motion in some given mode using a time-stepping method until an event is detected, either by some external signal (a discrete input, such as the turning of a switch) or by violation of some constraints on the continuous state. If such an event occurs, a search is made to find accurately the time of the event and the corresponding state values, and then the integration is restarted from the new initial time and initial condition in the "correct" mode; possibly a search has to be performed to find the correct mode.

In general one should expect that the behavior within a given mode is actually given by a mixture of differential and algebraic equations; for instance in the simulation of a sliding mode one has such a situation, cf. Chapter 3. The numerical integration of systems of DAEs has received considerable attention in recent years, see for instance the books by Brenan *et al.* [28] and by Hairer and Wanner [61] for more information. In the context of hybrid systems, start-up procedures for DAE solvers should receive particular attention since re-initializations are expected to occur frequently.

Events within hybrid system simulation can be distinguished in what we called *externally induced events* and *internally induced events*. Externally induced events force a change of mode at a certain time known in advance, such as when switches are turned in an electrical network according to a predetermined schedule. Internally induced (or state) events are more difficult to handle; these are the events that occur for internal reasons, such as when an inequality constraint becomes active. To catch the internally induced events, a hybrid system simulator needs to be equipped with an event detection module. Such a module will monitor the sign of certain functions of the state to see if the required inequality constraints are still satisfied. In the combination with a time stepping algorithm for the simulation of continuous dynamics, one has to take into account that the time at which an event takes place will in general not coincide with one of the grid points that the continuous simulator has placed on the time axis. Both the event time itself and the value of the continuous state at the time of the event will have to be found by some interpolation method.

The problem of finding the next discrete state is called the *mode selection problem*. The problem may be easy in some cases, for instance in a system in which buffers are emptied in a fixed order. There are other cases though in which the problem can be quite complicated. Consider for instance a pile of boxes whose relative motion is subject to a Coulomb friction law. Suppose that support on one side is suddenly removed so that the pile will tumble under the force of gravity. It is nontrivial to determine which boxes will start sliding with respect to each other and which ones will not. In the case of linear complementarity systems it will be shown in Chapter 4 how the next mode can be actually determined by solving an algebraic problem, and one can use this as a starting point to look for efficient numerical methods. In cases in which the new mode has to be selected partly on the basis of information from the continuous state, one is dealing with a nonconstant mapping from a continuous domain to a discrete domain. Such a mapping can never be continuous and so one will have to live with the fact that in some cases decisions will be very sensitive. In such situations the simulation software should provide a warning to the user, and if it is difficult to make a definitive choice between several possibilities perhaps the solver should even work out all reasonable options in parallel.

In many hybrid systems the trajectories of continuous variables can be expected to be continuous functions, and in these cases the problem of re-initialization comes down to determining the value of the continuous state at the event time so that the simulation of the smooth dynamics in the new regime can start from an initial state that is correct up to the specified tolerance. In some cases however, such as in mechanical systems subject to unilateral constraints, jumps need to be calculated, see e.g. the examples provided in Chapter 2. Theoretically, the state after the jump should satisfy certain constraints exactly; finite word length effects however will cause small deviations in the order of the machine precision. Such deviations may cause an interaction with the mode selection module; in particular it may appear that a certain constraint is violated so that a new event is detected. In this way it may happen that a cycling between different modes occurs ("livelock"), and the simulator does not return to a situation in which motion according to some continuous dynamics is generated, so that effectively the simulation stops.

3. *The timestepping method.* In a number of papers (see for instance [111, 124, 146] it has been suggested that in fact it may not be necessary to track events in order to obtain approximate trajectories of hybrid systems. Moreover, such methods have already been used to implement simulators for demanding applications like the simulation of integrated circuits with thousands of transistors [93]. The term "timestepping methods" has been used to refer to methods that not aim to determine event times; we shall use this term as well, even though of course also the event-tracking methods use time discretization. Rather than giving a formal discussion of timestepping methods, let us illustrate the idea in an example.

Consider the following system, which is actually a time-reversed version of an example of Filippov [54, p. 116]. Let a relay system be given by

$$
\begin{aligned}
\dot{x}_1(t) &= -\operatorname{sgn} x_1(t) + 2\operatorname{sgn} x_2(t) \\
\dot{x}_2(t) &= -2\operatorname{sgn} x_1(t) - \operatorname{sgn} x_2(t)
\end{aligned} \tag{1.12}
$$

where the signum function (or relay element) sgn is actually not a function but a relation (or multi-valued function) specified by

$$
(x > 0 \,\wedge\, \operatorname{sgn} x = 1) \,\vee\, (x < 0 \,\wedge\, \operatorname{sgn} x = -1) \,\vee
$$
$$
(x = 0 \,\wedge\, -1 \leq \operatorname{sgn} x \leq 1). \tag{1.13}
$$

The system (1.12) may be described as a *piecewise constant* system; in each quadrant of the (x_1, x_2)-plane the right hand side is a constant vector. The interpretation of systems containing relay elements will be further discussed in Chapter 3. In the case of the simple example above it is strongly suggested that solutions should be as pictured in Fig. 1.4. The solutions are spiraling towards the origin, which is an equilibrium point. It can be verified that $\frac{d}{dt}(|x_1(t)| + |x_2(t)|) = -2$ which means that solutions starting at (x_{10}, x_{20}) cannot stay away from the origin for longer than $\frac{1}{2}(|x_{10}| + |x_{20}|)$ units of time. However, solutions cannot arrive at the origin without going through an infinite number of mode switches; since these mode switches would have to occur in a finite time interval, there must be an accumulation of events.

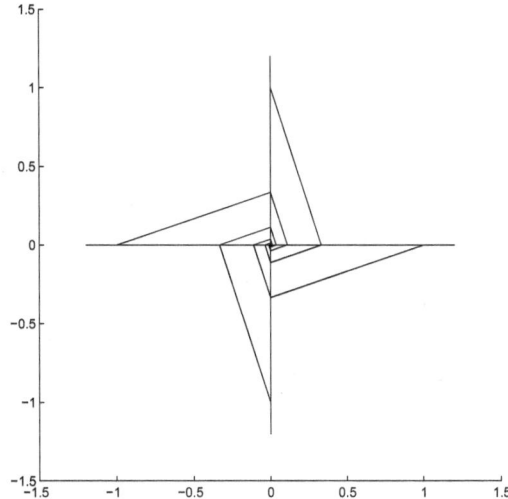

Figure 1.4: Solutions of Filippov's example (1.12)

Clearly an event-tracking method is in principle not able to carry out simulation across the accumulation point. The simplest fixed-step discretization scheme for (1.12) is the forward Euler scheme

$$
\begin{aligned}
\frac{x_{1,k+1} - x_{1,k}}{h} &= -\operatorname{sgn} x_{1,k} + 2\operatorname{sgn} x_{2,k} \\
\frac{x_{2,k+1} - x_{2,k}}{h} &= -2\operatorname{sgn} x_{1,k} - \operatorname{sgn} x_{2,k}
\end{aligned}
\tag{1.14}
$$

where h denotes the size of the time step and the variable $x_{i,k}$ ($i = 1,2$) is intended to be an approximation to $x_i(t)$ for $t = kh$. With the interpretation (1.13) of the signum function this discrete-time system is not deterministic, however. An alternative is to use an implicit scheme. The simplest choice of such a scheme is the following:

$$
\begin{aligned}
\frac{x_{1,k+1} - x_{1,k}}{h} &= -\operatorname{sgn} x_{1,k+1} + 2\operatorname{sgn} x_{2,k+1} \\
\frac{x_{2,k+1} - x_{2,k}}{h} &= -2\operatorname{sgn} x_{1,k+1} - \operatorname{sgn} x_{2,k+1}.
\end{aligned}
\tag{1.15}
$$

At each step, $x_{1,k}$ and $x_{2,k}$ are given and (1.15) is to be solved for $x_{1,k+1}$ and $x_{2,k+1}$. The equations (1.15) may be written as a system of equalities and inequalities by introducing some extra variables. Simplifying notation a bit by writing simply x_i instead of $x_{i,k+1}$ and $x_i^\flat$ for $x_{i,k}$, we obtain the following set of equations and inequalities:

$$
\begin{aligned}
x_1 &= x_1^\flat - hu_1 + 2hu_2 \tag{1.16a} \\
x_2 &= x_2^\flat - 2hu_1 - hu_2 \tag{1.16b}
\end{aligned}
$$

$$(x_1 > 0 \wedge u_1 = 1) \vee (x_1 < 0 \wedge u_1 = -1) \vee (x_1 = 0 \wedge -1 \leq u_1 \leq 1)$$
$$\text{(1.16c)}$$

$$(x_2 > 0 \wedge u_2 = 1) \vee (x_2 < 0 \wedge u_2 = -1) \vee (x_2 = 0 \wedge -1 \leq u_2 \leq 1).$$
$$\text{(1.16d)}$$

This system is to be solved in the unknowns x_1, x_2, u_1, and u_2 for arbitrary given $x_1^\flat$ and $x_2^\flat$; h is a parameter. It can be verified directly that for each positive value of h and for each given $(x_1^\flat, x_2^\flat)$ the above system has a unique solution; alternatively, one may recognize the system (1.16) as an instance of the Linear Complementarity Problem of mathematical programming and infer the same result from general facts about the LCP. In Figure (1.5) we show the partitioning of the $(x_1^\flat, x_2^\flat)$ plane that corresponds to the nine possible ways in which the disjunctions in (1.16c–1.16d) can be satisfied. For instance, the solution that has $x_1 = 0$ and $x_2 = 0$ is obtained for the values of $(x_1^\flat, x_2^\flat)$ such that the solution (u_1, u_2) of

$$\begin{bmatrix} x_1^\flat \\ x_2^\flat \end{bmatrix} + h \begin{bmatrix} -1 & 2 \\ -2 & -1 \end{bmatrix} \begin{bmatrix} u_1 \\ u_2 \end{bmatrix} = \begin{bmatrix} 0 \\ 0 \end{bmatrix} \tag{1.17}$$

satisfies $|u_1| \leq 1$ and $|u_2| \leq 1$. A simple matrix inversion shows that this happens when

$$-5h \leq x_1^\flat + 2x_2^\flat \leq 5h, \quad -5h \leq -2x_1^\flat + x_2^\flat \leq 5h \tag{1.18}$$

which corresponds to the central area in Fig. 1.5. The solution of the discretized system behaves like that of the original continuous system except in the narrow strips which do not influence the solution very much, and except in the central area where the discretized solution jumps to zero whereas the continuous system continues to go through mode changes at a higher and higher pace. Although we do not present a formal proof here, it is plausible from the figure that, when the step size h tends to zero, the solution of the discretized system converges to the solution of the original system, including the continuation of this solution by $x(t) = 0$ beyond the accumulation of event times. This happens in spite of the fact that the discretized system only goes through finitely many mode changes. Note also that the explicit scheme (1.14) shows a rather different and much less satisfactory behavior.

The discussion of the example suggests that at least in some cases and by using suitably selected discretization schemes it is possible to get an accurate approximation of the trajectories of a hybrid systems without tracking events. Obviously there are many questions to be asked, such as under what conditions it is possible to use a timestepping method, which discretization methods are most suitable, which orders of convergence can one get, and what can be gained by using a variable step size rather than a fixed step size. These issues are to a large extent a matter of future research.

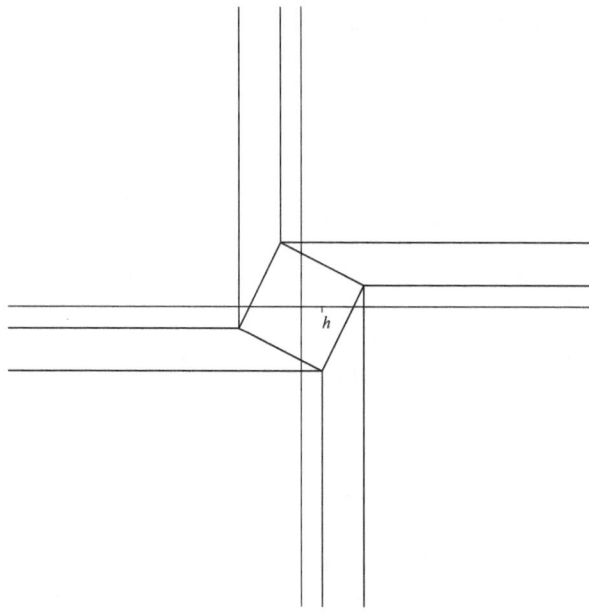

Figure 1.5: Partitioning of the plane induced by (1.15)

1.2.8 Representations of hybrid systems

In general, the quality and effectiveness of any mathematical model depends crucially on the purpose one wants to use it for. This is particularly true for models of complex systems, and thus for most hybrid systems. Furthermore, mathematical models of the same system, which are exactly equivalent, may have very different properties from the user's point of view. Also, the mathematical model (or language) describing the *functioning* of the system may not be the same as the appropriate language formulating the requirements which the system is expected to satisfy (the *specifications*). These features are particularly present in the field of discrete-event systems (or *distributed* or *concurrent* systems) where one finds in the literature a wealth of different descriptions, from process-algebraic formalisms such as CCS, CSP, ACP, LOTOS and logical theories such as temporal logic to graphically oriented approaches (net theories) such as Petri nets and automata, all with their own advantages and disadvantages.

As a result, it is to be expected that mathematically equivalent descriptions of a hybrid system, that is, different *representations* of the same hybrid system, have very different properties, depending on the purpose one wants to use it for. Also, it is clear that certain representations are better suited for treating a specific subclass of the wealth of hybrid systems than others. Finally, since the representation formalism often defines the starting point for the development of tools for automatically checking certain system properties, the resulting algorithmic properties of different representations may differ considerably.

In conclusion, different representations of hybrid systems have their own pros and cons, and one cannot hope for a single representation that will be suitable in all cases and for all purposes. Let us briefly and tentatively discuss some of the expected merits of the hybrid system representations we have seen so far.

It may seem clear that the *behavioral definition* of a hybrid system (Definition 1.2.6) is useful from a conceptual point of view and for theoretical purposes, but for most other purposes will not yield a convenient description. Instead, in order to carry out algorithms, one will usually need a more manageable and compact hybrid system representation.

Definitions 1.2.3 and 1.2.5 of a (generalized) hybrid automaton do provide workable representations of hybrid systems for various aims. First of all, they offer a clear picture of hybrid dynamics, which is very useful for exposition and theoretical analysis. A favorable feature of the hybrid automaton model is that the *semantics* of the model is quite explicit, as we have seen above. Furthermore, for a certain type of hybrid systems and for certain applications, the hybrid automaton representation can be quite effective.

Nevertheless, a drawback of the hybrid automaton representation is its tendency to become rather complicated, as we shall see already in some of the examples provided in the next chapter. This is foremost due to the fact that in the hybrid automaton model it is necessary to specify all the locations and all the transitions from one location to another, together with all their guards and jumps (or to completely specify the subset R of Definition 1.2.5). If the number of locations grows, this usually becomes an enormous and error-prone task. Other related types of (graphical) representations of hybrid systems that have been proposed in the literature, such as differential (or dynamically colored) *Petri nets*, may be more efficient than the hybrid automaton model in certain cases but have similar features.

For hybrid systems arising in a *"physical domain"* it seems natural to use representation formalisms such as event-flow formulas, which are closer to first principles physical modeling. First principles modeling of dynamical systems almost invariably leads to sets of *equations*, differential or algebraic. Furthermore, the hybrid nature of such systems is usually in first instance described by "if-then" or "either-or" statements, in the sense that in one location of the hybrid system a particular subset of the total set of differential and algebraic equations has to be satisfied, while in another location a different subset of equations should hold. Thus, while in the (generalized) hybrid automaton model the dynamics associated with every location are in principle completely independent, in most "physical" examples (as we will see in Chapter 2) the *set of equations* describing the various activities or *modes* (continuous dynamics associated to the locations) will remain almost the same, replacing one or more equations by some others.

Seen from this perspective, the (generalized) hybrid automaton model (and other similar descriptions of hybrid systems) may be quite far from the kind of model one obtains from physical first principles modeling, and the translation of the modeling information provided by equations, inequalities and logical statements into a complete specification of all the locations of the hybrid automaton together with all the possible discrete transitions and the complete continuous-time dynamics of every location may be a very tedious operation for the user. This becomes especially clear in an object-oriented modeling approach, where the interconnection (or composition) of hybrid automata may easily lead to a rapid growth in number of locations, and a rather elaborate (re-)specification of the resulting hybrid automaton model obtained by interconnection. Thus from the user's point of view an interesting alternative for efficiently specifying "physical" hybrid systems is to look for possibilities of specifying such systems primarily *by means of equations*, as in the framework of event-flow formulas. The setting of event-flow formulas is close to that of simulation languages such as Modelica™ [50]. Some of the modeling constructs in Modelica relating to hybrid systems do in fact have the form of event-flow formulas. Synchronous languages like LUSTRE [62] and SIGNAL [19] are also related, be it more distantly since these languages operate in discrete time; see [18] for an approach to general hybrid systems inspired by the SIGNAL language.

The formalism of event-flow formulas results in rather *implicit* representations of a hybrid system, as opposed to the almost completely *explicit* representations provided by the (generalized) hybrid automaton model. The price that has to be paid for the use of more implicit representations is that some of the problems in specifying the hybrid system are shifted to the definition of its solutions (the *semantics*).

Within the framework of event-flow formulas one still strives for *complete* specifications of the hybrid system under consideration. In some examples (e.g. the two-carts example, the power-converter example and the variable-structure systems example in Chapter 2) the initial description of the hybrid system obtained from first principles modeling is *incomplete*, especially with regard to the specification of the discrete dynamics. In fact, one would like to *automatically* generate a complete event-flow formula description based on this initial, incomplete, description, together with some additional information, like the assumption of elastic or non-elastic collisions in Subsection 2.2.9. In Chapter 4 we will in fact work out such a framework for a special class of hybrid systems, called *complementarity* hybrid systems (including some of the examples given in Chapter 2).

A wealth of different formalisms for describing hybrid systems have been proposed and are beginning to emerge in the literature. Most of them are extensions of formalisms for describing concurrent systems (extended duration calculus, hybrid CSP, hybrid state charts, TLA+, Z and duration calculus, VDM++, etc.), and are efficient only for relatively simple continuous dynamics, such as clock time evolution ($\dot{t} = 1$ or $\dot{t} = c$, where c is some constant) or continuous dynamics which can be reasonably approximated in this way. From a general point of view it seems natural to try to combine process-algebraic for-

malisms with the description of continuous dynamics by differential-algebraic
equations; but no general theory has emerged so far in this direction.

1.3   Notes and References for Chapter 1

A broad view on the research activities in the area of hybrid systems during the
last decade can be obtained from the conference proceedings [58], [5], [3], [100],
[6], [137], [153], as well as the journal special issues [7] and [116]. Needless to
say that many valuable aspects and/or approaches covered in the literature
will not be addressed in the presentation of hybrid systems in this text. The
terminology "hybrid systems" for this class of systems seems to have been first
used by Witsenhausen [159]. A recent introduction to various approaches in
the theory of *concurrent processes*, in particular CSP, can be found in [133];
see this book for further references. For temporal logic we refer to [104] and
[105]. For an interesting discussion on the similarities and differences between
the view points of on the one hand computer science and discrete event systems
(automata) and on the other hand systems and control theory we refer to [101].

There are several useful approaches to hybrid systems that we have not discussed here. Often these approaches have been developed with an eye towards
specific applications or techniques. We mention two directions in particular.
In the study of discrete-event systems, Petri nets enjoy great popularity because many situations can be modeled much more efficiently by a Petri net
than by a finite automaton with no special structure. For proposals concerning
extensions of Petri nets with continuous dynamics, see for instance [43, 119].
In contrast, the approach based on mixed logical dynamical (MLD) systems
introduced in [16] is a discrete extension of a continuous framework. In this
approach a class of hybrid systems is described by linear dynamic equations
subject to linear inequalities, on the basis of the correspondence that can be
constructed between propositional logic and linear inequalities in real and integer variables (see e.g. [157]).

Chapter 2

Examples of hybrid dynamical systems

2.1 Introduction

In this chapter we treat various (rather simple) examples of hybrid systems from different application areas, with the aim of illustrating the notions of hybrid models and dynamics discussed in the previous chapter. Some of the examples will also return in the developments of the following chapters.

In order to formalize the examples as hybrid systems we primarily use the notion of a hybrid automaton (Definition 1.2.3), or the framework of *event-flow formulas* as introduced in Subsection 1.2.6 of the previous chapter. A number of notational conventions will be used to facilitate the presentation of event-flow formulas; these have partly already been mentioned in Subsection 1.2.6. A *comma* is used to indicate logical conjunction between several expressions on one line, a *vertical bar* is used to indicate logical disjunction between several expressions on one line or between successive lines, and a *left curly bracket* is used to indicate logical conjunction between successive lines. Furthermore, $\|$ indicates parallel composition between two subsystems. The symbols associated with a given subsystem are not listed explicitly but are understood as being the symbols that occur in the Boolean expressions in the EFF for that subsystem. Variables are understood to be continuous across events by default, so we do not explicitly write conditions of the type $x_i^+ = x_i^-$; likewise, conditions of the form $S = \texttt{blank}$ are not written explicitly. Also the empty event that goes with each subsystem is not written explicitly.

2.2 Examples

2.2.1 Hysteresis

Consider a control system with a hysteresis element in the feedback loop (cf. [26]):

$$\dot{x} = H(x) + u \tag{2.1}$$

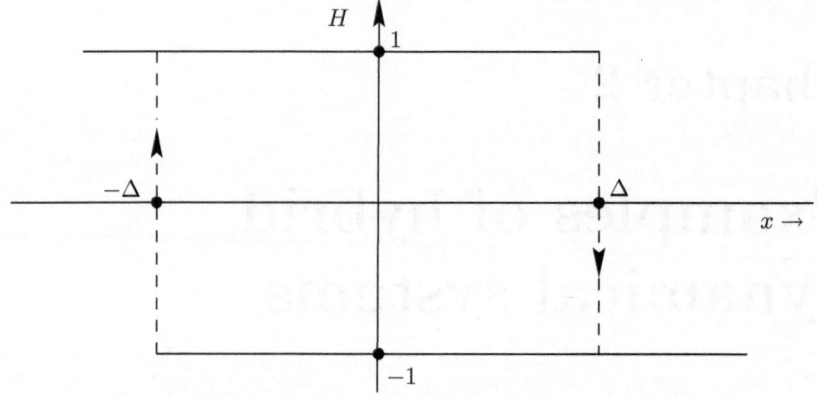

Figure 2.1: Hysteresis

where the multi-valued function H is shown in Figure 2.1. Note that this system is not just a differential equation whose right-hand side is piecewise continuous. There is "memory" in the system, which affects the right-hand side of the differential equation. Indeed, the hysteresis function H has an automaton naturally associated to it, and the system (2.1) can be formalized as the hybrid automaton depicted in Figure 2.2.

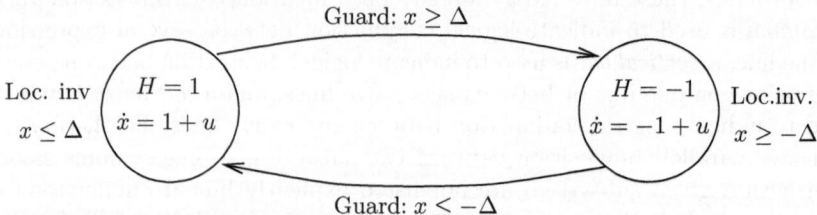

Figure 2.2: Control system with hysteresis as a hybrid automaton

In Figure 2.2 the location invariants are written inside the two circles representing the two locations, and the transitions (events) are labeled with their guards. Note that this is a hybrid system involving internally induced switchings, but no jumps.

If we follow the "persistent-mode convention" of Remark 1.2.22, then an event-flow formula for this system is simply given by

$$\begin{vmatrix} \dot{x} = 1 + u, & x \leq \Delta \\ \dot{x} = -1 + u, & x \geq -\Delta. \end{vmatrix} \tag{2.2}$$

2.2.2 Manual transmission

Consider a simple model of a manual transmission [30]:

$$\dot{x}_1 = x_2$$

$$\dot{x}_2 = \frac{-a.x_2 + u}{1 + v}$$

where v is the gear shift position $v \in \{1, 2, 3, 4\}$, u is the acceleration and a is a parameter of the system. Clearly, this is a hybrid system having four locations and a two-dimensional continuous state, with controlled transitions (switchings) and no jumps.

2.2.3 Bouncing ball

Consider a ball bouncing on a table; the bounces are modeled as being instantaneous, with restitution coefficient e assumed to be in the open interval $(0, 1)$. There are no discrete variables (there is only one location), and there is one continuous variable, denoted by q; this variable indicates the distance between the table and the ball.

In the hybrid automaton model of this system the system switches from the single location back to the same location while an (autonomous) *jump* occurs in the continuous state given by the position q and the velocity $\dot{q}$, since the velocity changes at an event (impact) time t from $\dot{q}(t^{0\sharp})$ into $\dot{q}(t^{\sharp}) = -e\dot{q}(t^{0\sharp})$. The guard of this transition (event) is given by $q = 0$, $\dot{q} \leq 0$.

The dynamics of the system can be summarized by the differential equation (after normalization of all constants)

$$\ddot{q} = -1 \tag{2.3}$$

if $q > 0$, together with the discrete transition (impact rule) at an event time τ

$$\dot{q}(\tau^{\sharp}) = -e\dot{q}(\tau^{0\sharp}), \tag{2.4}$$

which occurs if $q = 0$ and $\dot{q} \leq 0$

Furthermore the connection between the dynamics at "ordinary" time instants and the discrete transition at an event time τ is provided by the compatibility conditions

$$\lim_{t \uparrow \tau} q(t) = q(\tau^{0\sharp}) = q(\tau^{\sharp}) = \lim_{t \downarrow \tau} q(t)$$

$$\lim_{t \uparrow \tau} \dot{q}(t) = \dot{q}(\tau^{0\sharp}) \tag{2.5}$$

$$\dot{q}(\tau^{\sharp}) = \lim_{t \downarrow \tau} \dot{q}(t).$$

Using the convention that state variables are continuous across events unless indicated otherwise, one may write equations (2.3) and (2.4) in an alternative and more compact form as the event-flow formula (in the continuous state vector $x = (q, \dot{q})$)

$$
\begin{vmatrix}
q \geq 0, & \ddot{q} = -1 \\
q = 0, & \dot{q} = 0 \\
q = 0, & \dot{q}^+ = -e\dot{q}^-.
\end{vmatrix}
\tag{2.6}
$$

It is clear that the system can be consistently initialized by prescribing $q(0^-)$ and $\dot{q}(0^-)$, with $q(0^-) \geq 0$. In general it can be a nontrivial question for an event-flow formula to determine the data one has to provide at 0^- in order to ensure the existence of a unique solution with these initial data. Prescribing data at 0^- rather than at 0 allows 0 to be an event time.

In this example, event times must actually have accumulation points; for instance if we set $q(0^-) = 0$ and $\dot{q}(0^-) = 1$, then it is easily verified that bounces take place at times 2, $2 + 2e$, $2 + 2e + 2e^2, \ldots$, so that we have an accumulation point at $\frac{2}{1-e}$. Nevertheless we are still able to define a solution:

$$
\begin{aligned}
\mathcal{E}_T &= \{2 \sum_{j=0}^{k-1} e^j \mid k \in \mathbb{N}\} \cup \{\tfrac{2}{1-e}\} \\
q(t) &= e^k t - \tfrac{1}{2}(t - 2 \sum_{j=0}^{k-1} e^j)^2 && \text{for } t \in (2 \sum_{j=0}^{k-1} e^j, 2 \sum_{j=0}^{k} e^j), \\
&&& k = 0, 1, 2, \ldots \\
&= 0 && \text{for } t > \tfrac{2}{1-e}
\end{aligned}
\tag{2.7}
$$

(we have used the standard convention that a summation over an empty set produces zero). One easily verifies that this is the only piecewise differentiable solution to (2.6) that is continuous in the sense that the left- and right-hand limits of $q(t)$ exist and are equal to each other for all t. Note that in the present example the hybrid trajectory can be naturally extended *after* the accumulation point $t = 2(1 - e)^{-1}$ of event times, since both the continuous state and the discrete state (there is only one!) converge at the accumulation point. In the present case the extension is just the zero trajectory, but it is easy to modify the example in such a way that the extension is more involved. For some examples of hybrid systems having more than one location, where the existence of an extension after the accumulation point is more problematic, we refer to [82].

2.2.4   Temperature control system

In a simple model to be used in temperature control of a room (cf. [1]), we have one continuous variable (room temperature, denoted by $\theta(t)$, and taking values in $\mathbb{R}$) and one discrete variable (status of the heater, denoted by $H(t)$ and taking values in $\{\textsf{on}, \textsf{off}\}$). The continuous dynamics in the system may be described by an equation of the form

$$\dot{x} = f(x, w, H), \quad \theta = g(x) \tag{2.8}$$

where $f(\cdot, \cdot)$ is a (sufficiently smooth) function of the continuous state x, the discrete state H, and the continuous external variable w, which may for example contain the outside temperature. The hybrid automaton model of this system is given by two locations with the obvious location invariants, as given in Figure 2.3. Note that the specification of the guards is crucial for the set of

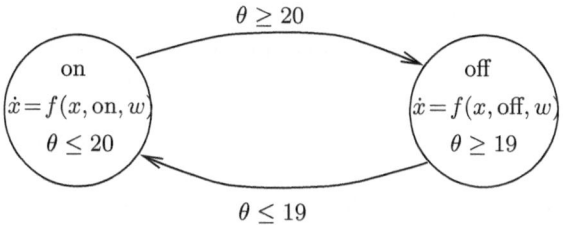

Figure 2.3: Temperature-thermostat system as a hybrid automaton

trajectories of the hybrid automaton. For example, if we take the guards to be $\theta \geq 19$, respectively $\theta \leq 20$, then transition from one location to another is permitted while the temperature θ is between 19 and 20, and the temperature-thermostat system will behave non-deterministically. This specification may be appropriate in particular when the thermostat is actually a human controller, or in other circumstances in which the thermostat takes other factors into consideration than just the one that is explicitly specified in the model. A more detailed specification may not be worth the effort for some applications. A more restrictive specification of the guards would be $\theta = 20$, respectively $\theta = 19$; note however that this is still not quite enough to obtain unique state trajectories for all input trajectories.

In an event-flow description the system is explicitly modeled as a composition of two subsystems, namely the heater and the thermostat. So we write

$$\text{system} = \text{heater} \parallel \text{thermostat} \tag{2.9}$$

where the subsystem "heater" is given by (2.8) in which H now acts as a discrete communication variable, and the subsystem "thermostat" is given by

$$\left|\begin{array}{ll} \theta \leq 20, & H = \texttt{on} \\ \theta \geq 19, & H = \texttt{off} \\ \theta \leq 20, & H^{+} = \texttt{on} \\ \theta \geq 19, & H^{+} = \texttt{off}. \end{array}\right. \tag{2.10}$$

If it is desired to specify for instance that the heating can only be turned on when the temperature is exactly 19 degrees, then the first event condition in (2.10) should be replaced by

$$\theta = 19, \quad H^- = \mathtt{off}, \quad H^+ = \mathtt{on}. \tag{2.11}$$

Alternatively we can impose the "persistent-mode convention" discussed in Remark 1.2.22.

2.2.5 Water-level monitor

Our next example (taken from [1]) is somewhat more elaborate than the ones that we discussed before. The example concerns the modeling of a water-level control system. There are two continuous variables, denoted by $y(t)$ (the water level) and $x(t)$ (time elapsed since last signal was sent by the monitor). There are also two discrete variables, denoted by $P(t)$ (the status of the pump, taking values in $\{\mathtt{on}, \mathtt{off}\}$) and $S(t)$ (the nature of the signal last sent by the monitor, also taking values in $\{\mathtt{on}, \mathtt{off}\}$). The dynamics of the system is given in [1] as follows. The water level rises one unit per second when the pump is on and falls two units per second when the pump is off. When the water level rises to 10 units, the monitor sends a switch-off signal, which after a delay of two seconds results in the pump turning off. When the water level falls to 5 units, the monitor sends a switch-on signal, which after a delay of again two seconds causes the pump to switch on.

There are several ways in which one may write down equations to describe the system, which may be related to various ways in which the controller may be implemented. For instance the monitor should send a switch-off signal when the water level reaches 10 units and is rising, but not when the level reaches 10 units on its way down. This may be implemented by the sign of the derivative of y, by looking at the status of the pump, or by looking at the signal last sent by the monitor. Under the assumptions of the model these methods are all equivalent in the sense that they produce the same behavior; however there can be differences in robustness with respect to unmodeled effects. The solution proposed in [1] is based on the signal last sent by the monitor. The hybrid automaton model of this system is given in Figure 2.4.

For a description by means of event-flow formulas it seems natural to use parallel composition. One has to spell out in which way the monitor knows whether the water level is rising or falling when one of the critical levels is observed. Here we shall assume that the monitor remembers which signal it has last sent. For that purpose the monitor needs to have a discrete state variable. We can then write the system as follows:

$$\text{system} \; = \; \text{tank} \,\|\, \text{pump} \,\|\, \text{monitor} \,\|\, \text{delay} \tag{2.12a}$$

with

$$\text{tank} \left| \begin{array}{ll} P = \mathtt{on}, & \dot{y} = 1 \\[2mm] P = \mathtt{off}, & \dot{y} = -2 \end{array} \right. \tag{2.12b}$$

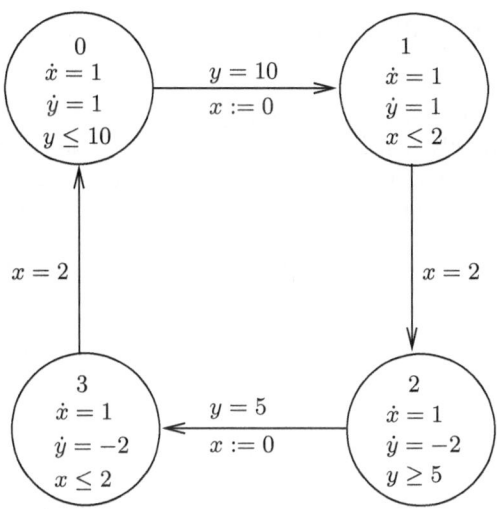

Figure 2.4: Water-level monitor

$$\text{pump} \left|\begin{array}{ll} S' = \text{sw\_on}, & P^+ = \text{on} \\ S' = \text{sw\_off}, & P^+ = \text{off} \end{array}\right. \tag{2.12c}$$

$$\text{monitor} \left|\begin{array}{lll} y \le 10, & Q = \text{req\_off} \\ y \ge 5, & Q = \text{req\_on} \\ y = 10, & Q^- = \text{req\_on}, & Q^+ = \text{req\_off}, & S = \text{sw\_off} \\ y = 5, & Q^- = \text{req\_off}, & Q^+ = \text{req\_on}, & S = \text{sw\_on} \end{array}\right.$$
$$\tag{2.12d}$$

$$\text{delay} \left|\begin{array}{ll} D = \text{inactive}, & \dot{\tau} = 0 \\ D \ne \text{inactive}, & \dot{\tau} = 1, \quad \tau \le 2 \\ D^+ = S, & \tau^+ = 0 \\ \tau = 2, & S' = D^-, \quad \tau^+ = 0, \quad D^+ = \text{inactive}. \end{array}\right. \tag{2.12e}$$

Remark 2.2.1. When active, the delay needs a clock (implemented by the variable τ) in order to tell when two units of time have passed. When the delay is inactive it doesn't need the clock, however; therefore it would perhaps be better to say that during these periods the clock is "nonexistent" rather than to give it some arbitrary dynamics. This would require a modification of the setting described here in the spirit of the "presences" of [18].

2.2.6   Multiple collisions

The area of mechanical collisions offers a wealth of examples for hybrid systems modeling, see e.g. [31] and the references quoted therein. We will only consider the following seemingly simple case, which is sometimes known as Newton's cradle (with three balls). Consider three point masses with unit mass, moving on a straight line. The positions of the three masses will be denoted by q_1, q_2, q_3, with $q_1 \leq q_2 \leq q_3$. When no collisions take place the dynamics of the three masses is described by the second-order differential equations

$$\dot{q}_i = v_i$$
$$\dot{v}_i = F_i$$

for $i = 1, 2, 3$, with F_i some prespecified force functions (e.g., gravitational forces).

Collisions (events) take place whenever the positions q_i and the velocities v_i satisfy relations of the form

$$q_i = q_j$$
$$v_i > v_j$$

for some $i \in \{1, 2, 3\}$ with $j = i + 1$.

The usual impact rule for *two* rigid bodies a and b with masses m_a, respectively m_b, specifying the velocities $v_a^\sharp, v_b^\sharp$ after the impact, is given by

$$v_a^\sharp - v_b^\sharp = -e(v_a - v_a) \tag{2.13a}$$
$$m_a v_a^\sharp + m_b v_b^\sharp = m_a v_a + m_b v_b \tag{2.13b}$$

where $e \in [0, 1]$ denotes the restitution coefficient. Of course, the impact rule (2.13) constitutes already an event idealization of the "real" physical collision phenomenon, which includes a fast dynamical transition phase. Also note that (2.13b) expresses conservation of momentum for *simple* impacts. Below we shall apply this rule in a situation where multiple impact occurs (as it was done in [63]); of course one may debate whether such an extension is "correct".

Let us now consider Newton's cradle with three balls, and suppose that at some given time instant $t \in \mathbb{R}$ we have

$$q_1 = q_2 = q_3 = 0$$
$$v_1 = 1 \tag{2.14}$$
$$v_2 = v_3 = 0.$$

How do we model this case with the above impact rule? The problem is that not only the left mass is colliding with the middle mass, but that also the middle mass is in contact with the right mass; we thus encounter a *multiple* collision.

There are at least two ways to model this multiple collision based on the impact rule (2.13) for single collisions, and these two ways lead to different answers! One option is to regard the middle and the right mass at the moment of collision as a *single* mass, with mass equal to 2. Application of the impact rule (2.13) then yields

$$
\begin{aligned}
v_1^+ &= \tfrac{1}{3}(1 - 2e) \\
v_c^+ &= \tfrac{1}{3}(1 + e)
\end{aligned}
$$

where v_c denotes the velocity of the (combined) middle and right mass. For $e = 1$ (perfectly elastic collision) this yields the outcome $v_1^+ = -\tfrac{1}{3}, v_c^+ = \tfrac{2}{3}$, while for $e = 0$ (perfectly inelastic collision) we obtain the outcome $v_1^+ = v_c^+ = \tfrac{1}{3}$. Another way of modeling the multiple impact based on the single impact rule (2.13) (cf. [63]) is to imagine that the collision of the left mass with the middle mass takes place *just before* the collision of the middle mass with the right mass, leading to an event with multiplicity at least equal to 2.

For $e = 1$ this alternative modeling leads to the following different description. The impact rule (2.13) specializes for $e = 1$ to the event-clause

$$
\begin{aligned}
v_i^\# &= v_j \\
v_j^\# &= v_i.
\end{aligned}
$$

Hence, we obtain for the initial condition (2.14) an event with multiplicity 2, representing the transfer of the velocity $v_1 = 1$ of the first mass to the second and then to the third mass, with the velocities of the first and second mass being equal to zero after the collision. This behavior is definitely different from the behavior for $e = 1$ derived above, but seems to be reasonably close, at least for small time, to what one observes experimentally for "Newton's cradle".

On the other hand, for $e = 0$ (perfectly inelastic collision), we obtain in the second approach from (2.13) the event clause

$$
v_i^\# = \frac{1}{2}(v_i + v_j) = v_j^\#.
$$

With the same initial conditions (2.14) as above, this gives rise to an event with multiplicity equal to ∞. In fact, we obtain the following distribution of velocities at the subsequents stages $t^{0\#}, t^\#, t^{\#\#}, t^{\#\#\#}, \ldots$ of the time event at time t:

	v_1	v_2	v_3
$t^{0\#}$	1	0	0
$t^\#$	$\frac{1}{2}$	$\frac{1}{2}$	0
$t^{\#\#}$	$\frac{1}{2}$	$\frac{1}{4}$	$\frac{1}{4}$

$t^{\#\#\#}$	$\frac{3}{8}$	$\frac{3}{8}$	$\frac{1}{4}$
	$\downarrow$	$\downarrow$	$\downarrow$
$t^{\infty\#}$	$\frac{1}{3}$	$\frac{1}{3}$	$\frac{1}{3}$

Hence the outcome of this event with multiplicity ∞ is the same as the outcome derived by the first method, with an event of multiplicity 1.

Interesting variations of both modeling approaches may be obtained by considering e. g. four unit masses, with initial conditions

$$q_1 = q_2 = q_3 = q_4 = 0$$
$$v_1 = v_2 = 1 \tag{2.15}$$
$$v_3 = v_4 = 0.$$

Another interesting issue which may be studied in the context of this example concerns the continuous dependence of solutions on initial conditions. For a brief discussion of this subject in an example with inelastic collisions, see also Remark 4.5.2 below.

2.2.7 Variable-structure system

Consider a control system described by equations of the form $\dot{x}(t) = f(x(t), u(t))$, where $u(t)$ is the scalar control input. Suppose that a switching control scheme is employed that uses a state feedback law $u(t) = \phi_1(x(t))$ when the scalar variable $y(t)$ defined by $y(t) = h(x(t))$ is positive and a feedback $u(t) \doteq \phi_2(x(t))$ when $y(t)$ is negative. Writing $f_i(x) = f(x, \phi_i(x))$ for $i = 1, 2$, we obtain the dynamical system

$$\dot{x} = f_1(x) \quad \text{if} \quad h(x) \geq 0$$
$$\dot{x} = f_2(x) \quad \text{if} \quad h(x) \leq 0. \tag{2.16}$$

Such a system is sometimes called a *variable-structure system*. The precise interpretation of the above equations, which are in principle ambiguous since there is no requirement that $f_1(x) = f_2(x)$ when $h(x) = 0$, will be discussed briefly here and more extensively in Chapter 3.

A variable-structure system can be considered as a hybrid system with two locations having different activities; the expressions $h(x) \geq 0$ and $h(x) \leq 0$ serve as location invariants. The problem in specifying this hybrid system is to define the trajectories of the system, starting from initial conditions on the surface $h(x) = 0$. The combined vector field $f(x)$ defined by $f(x) := f_1(x)$ for $h(x) > 0$ and $f(x) := f_2(x)$ for $h(x) < 0$ is in general discontinuous on the switching surface $h(x) = 0$. Hence the standard theory for existence and uniqueness of solutions of differential equations does not apply, and, indeed, it

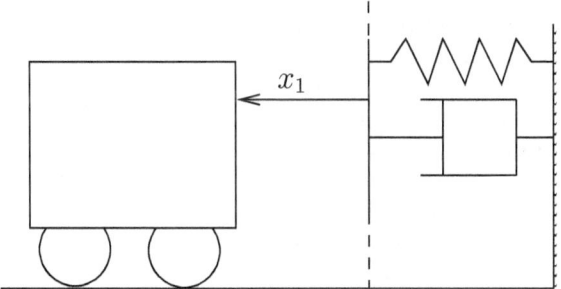

Figure 2.5: Collision to an elastic wall

is easy to come up with examples exhibiting multiple solutions for the continuous state x. Furthermore, it cannot be expected that state trajectories $x(t)$ are differentiable at points where the switching surface is crossed, so it would be too much to require that solutions satisfy (2.16) for all t. One possible way out is to replace the equation (2.16) by the integral form

$$x(t) = x(0) + \int_0^t f(x(s))ds \qquad (2.17)$$

which doesn't require the trajectory $x(\cdot)$ to be differentiable. A solution of (2.17) is called a solution of (2.16) in the sense of *Carathéodory*. The interpretation (2.17) also obviates the need for specifying the value of f on the surface $\{x \mid h(x) = 0\}$, at least for cases in which solutions arrive at this surface from one side and leave it immediately on the other side.

As an example consider the following model of an elastic collision, taken from [80]. Consider a mass colliding with an elastic wall; the elasticity of the wall is modeled as a (linear) spring-damper system, as shown in Figure 2.5. The system is described as a system with two locations (or modes):

$$\text{mode 0:} \quad \begin{cases} \begin{bmatrix} \dot{x}_1 \\ \dot{x}_2 \end{bmatrix} = \begin{bmatrix} 0 & 1 \\ 0 & 0 \end{bmatrix} \begin{bmatrix} x_1 \\ x_2 \end{bmatrix} + \begin{bmatrix} 0 \\ 1 \end{bmatrix} u \\[2em] y = \begin{bmatrix} 1 & 0 \end{bmatrix} \begin{bmatrix} x_1 \\ x_2 \end{bmatrix} \geq 0 \end{cases}$$

$$\text{mode 1:} \quad \begin{cases} \begin{bmatrix} \dot{x}_1 \\ \dot{x}_2 \end{bmatrix} = \begin{bmatrix} 0 & 1 \\ -k & -d \end{bmatrix} \begin{bmatrix} x_1 \\ x_2 \end{bmatrix} + \begin{bmatrix} 0 \\ 1 \end{bmatrix} u \\[2em] y = \begin{bmatrix} 1 & 0 \end{bmatrix} \begin{bmatrix} x_1 \\ x_2 \end{bmatrix} \leq 0 \end{cases}$$

Note that for $d \neq 0$ the overall dynamics is not continuous on the surface $y = x_1 = 0$, so that the standard theory of existence and uniqueness of solutions of differential equations does not apply. Nevertheless, it can be readily checked that the system has unique solutions for every initial condition (as we expect). On the other hand, modification of the equations may easily lead to a hybrid system exhibiting multiple solutions. Necessary and sufficient conditions for uniqueness of solutions in the sense of Carathéodory have been derived in [80, 81].

If in the general formulation the vector $f_1(x_0)$ points inside the set $\{x \mid h(x) < 0\}$ and the vector $f_2(x_0)$ points inside $\{x \mid h(x) > 0\}$ for a certain x_0 on the switching surface $h(x) = 0$, then clearly there does not exist a solution in the sense of Carathéodory. For this case another solution concept has been defined by Filippov, by *averaging* in a certain sense the "chattering behavior" around the switching surface $\{x \mid h(x) = 0\}$. From a hybrid systems point of view this can be interpreted as the creation of a new location whose continuous-time dynamics is given by this averaged dynamics on the switching surface. See for further discussion Chapter 3.

2.2.8   Supervisor model

The following model has been proposed for controlling a continuous-time input-state-output system (1.2) by means of an (input-output) finite automaton (see Definition 1.2.2); see [8, 120, 26] for further discussion. The model consists of three basic parts: continuous-time plant, finite control automaton, and interface. The interface in turn consist of two parts, viz. an analog-to-digital (AD) converter and digital-to-analog (DA) converter. The supervisor model is illustrated in Figure 2.6.

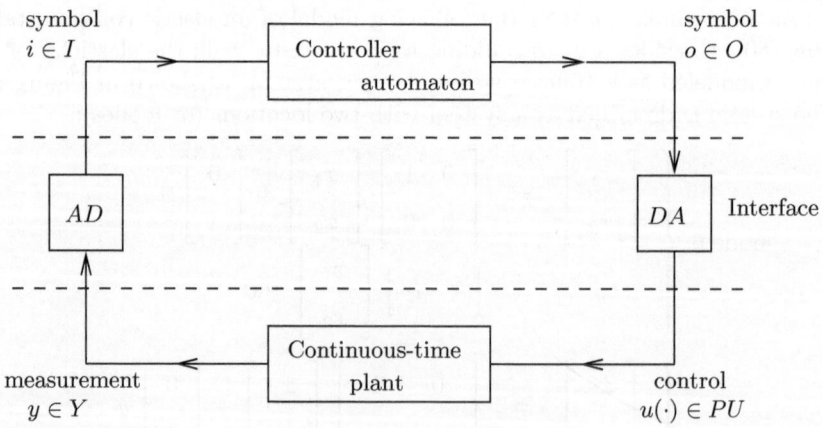

Figure 2.6: Supervisor model

Associated to the plant are an input space U, a state space X, and an output space Y, while the controller automaton has a (finite) input space I, a state space Q and an output space O. The controller automaton may be given as an input-output automaton as in (1.3)

$$q^\sharp = \nu(q, i)$$
$$o = \eta(q, i).$$

The AD converter is given by a map $AD : Y \times Q \to I$. The values of this map (the discrete input symbols) are determined by a partition of the output space Y, which may depend on the current value of the state of the finite automaton. The DA converter is given by a map $DA : O \to PU$, where PU denotes the set of piecewise right-continuous input functions for the plant.

The dynamics can be described as follows. Assume that the state of the plant is evolving and that the controller automaton is in state q. Then $AD(\cdot, q)$ assigns to output $y(t)$ a symbol from the input alphabet I. When this symbol changes, the controller automaton carries out the associated state transition, causing a corresponding change in the output symbol $o \in O$. Associated to this symbol is a control input $DA(o)$ that is applied as input to the plant until the input symbol of the controller automaton changes again.

In the literature the design of the supervisor is often based on a *quantization* (also sometimes called *abstraction*) of the continuous plant to a discrete event system. In this case one considers e.g. an appropriate (fixed) partition of the state space and the output space of the continuous time plant, together with a fixed set of input functions, and one constructs a (not necessarily deterministic) discrete-event system covering the quantized continuous-time dynamics. The events are then determined by the crossing of the boundaries defined by the partition of the state space.

2.2.9 Two carts

The logical disjunction (the "or" between propositions) historically has not been entirely absent from the study of continuous dynamical systems; in particular, disjunctions arise in the study of mechanical systems with unilateral constraints.

In the simplest case of a one-dimensional constraint, consider a nonnegative slack variable which is positive when the system is away from the constraint and zero when the system is at the constraint. The dynamics of the system will involve the disjunction of two possibilities: the slack variable is zero and the corresponding constraint force is nonnegative, or the constraint force is zero and the slack variable is nonnegative. This alternative occurs in classical textbooks on mechanics such as [125] and [86], and for the static case goes back to Fourier (cf. [92]). For a concrete example, consider the two-carts example that was discussed in [138] and [140]. Two carts are connected to each other and to a wall by springs; the motion of the left cart is constrained by a stop (see Fig. 2.7). It is assumed that the springs are linear, and all constants are set equal to 1; moreover, the stop is placed at the equilibrium position of the left cart. There are three continuous variables: $q_1(t)$, which denotes the

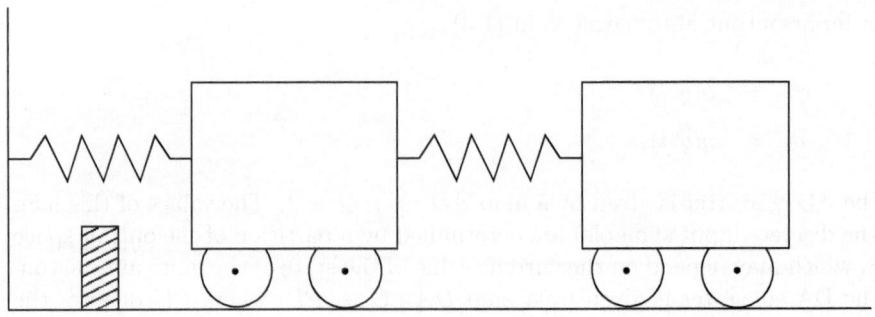

Figure 2.7: Two carts

position of the left cart measured with respect to its equilibrium position, so that $q_1(t)$ also serves as a slack variable; $q_2(t)$, which is the position of the right cart with respect to its equilibrium position; and $\lambda(t)$, which denotes the constraint force. The dynamics of the system can be succinctly written by the following event-flow formula, where $e \in [0, 1]$ is the restitution coefficient and the notation

$$x \perp y \quad :\Longleftrightarrow \quad \begin{vmatrix} x = 0 \\ y = 0 \end{vmatrix} \tag{2.18}$$

(for scalar variables x and y) is used:

$$\begin{vmatrix} \begin{cases} \ddot{q}_1 = -2q_1 + q_2 + \lambda \\ \ddot{q}_2 = q_1 - q_2 \\ 0 \le q_1 \perp \lambda \ge 0 \end{cases} \\ q_1 = 0, \quad \dot{q}_1^+ = -e\dot{q}_1^-. \end{vmatrix} \tag{2.19}$$

Classically the relation $0 \le q_1 \perp \lambda \ge 0$ is written as $q_1 \ge 0$, $\lambda \ge 0$, and $q_1\lambda = 0$, which indeed comes down to the same thing and avoids the explicit use of disjunctions.

The dynamics of the above system may also be given in a more explicit condition-event form, with the dynamics in each mode given in the form of ordinary differential equations, rather than differential-algebraic equations as in the mode descriptions that can be derived directly from (2.19). Since the differential equations are linear, they can even be solved explicitly so that one can obtain a full description as a hybrid system in the sense of [1]. This description however would be much longer than the one given above (see [141] where part of the description has been worked out). In general, a system

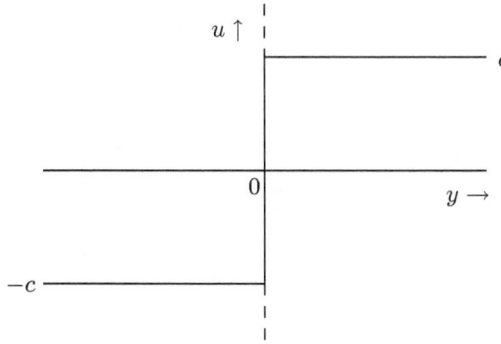

Figure 2.8: Coulomb friction characteristic

with k unilateral constraints can be described by k disjunctions as in (2.19), whereas the number of different locations (discrete states) following from the constraints is 2^k. In this sense the description by means of event-flow formulas represents an exponential saving in description length with respect to a formulation based on enumeration of the discrete states.

The trajectories $q_1(\cdot)$ may be taken as continuous and piecewise differentiable, but not as piecewise continuously differentiable since we have to allow jumps in the derivative. In particular, if the stop is assumed to be completely inelastic ($e = 0$) then the velocity of the left cart will be reduced to zero immediately when it hits the block.

In the present example, the specification of the jump rule (the event clause) is quite simple. However, in higher-dimensional examples of this type, especially when multiple collisions are involved, a complete a priori specification of the jumps may not be so easy. In fact, in many cases one would like to avoid an a priori complete specification of the event clauses, and an attractive alternative would be to automatically generate the event-clauses on the basis of inequality constraints of the type $q \geq 0$, $\lambda \geq 0$, and $q\lambda = 0$ as above, and physical information concerning the nature of the collisions, such as the restitition coefficient e (see also [31] for a discussion). In Chapter 4 we shall come back to this issue in the general context of *complementarity* hybrid systems, including the present example.

2.2.10   Coulomb friction

An element representing Coulomb friction can be constructed from two continuous variables, say $y(t)$ and $u(t)$, which are related in the following way (see Figure 2.8):

$$\text{Coulomb} \left|\begin{array}{ll} y \geq 0, & u = c \\ y = 0, & -c \leq u \leq c \\ y \leq 0, & u = -c. \end{array}\right. \qquad (2.20)$$

Here c is some non-negative constant, while $y(t)$ denotes the *velocity* of a mass and $u(t)$ denotes the *frictional force* applied to this mass, due to the Coulomb friction. The friction element above can be coupled to another system description (consisting for instance of a continuous system with inputs $u(t)$ (frictional forces) and outputs $y(t)$ (velocities) by conjunction. Whether or not the complete system will have solutions defined for arbitrarily large t depends on the nature of the system added. Notice also that it is not natural to assume a priori that the frictional forces $u(t)$ are continuous; imagine for example a heavy mass sliding subject to Coulomb friction along an upwards inclined plane, which after coming to rest will immediately slide down subject to the reversed frictional force. Hence we cannot a priori require the functions $u(t)$ and $y(t)$ appearing in (2.20) to be continuous. For the case in which the added system is a linear finite-dimensional time-invariant input-output system of the type usually studied in control theory, sufficient conditions for well-posedness (in the sense of existence and uniqueness of solutions) have been given in [95]; see also Chapter 4. The determination of the location in systems with multiple Coulomb friction elements is a nontrivial problem, cf. [129] and the comments in [50].

In many cases the constant c appearing in the Coulomb friction is a function of the *normal force* applied to the moving mass due to an imposed *geometric inequality constraint* of the type appearing in Subsection 2.2.9. In this way, by combining the characteristics of Subsection 2.2.9 with those of the present one, we can describe general *multi-body systems* with geometric inequality constraints and multiple Coulomb friction as hybrid systems; see e.g. [129, 31] for a more extensive discussion.

2.2.11 Systems with piecewise linear elements

Note that the Coulomb friction characteristic depicted in Figure 2.8 can be also interpreted as an ideal *relay element* (without deadzone). In this case, the third mode (or location) corresponding to the *vertical* segment of the characteristic is usually interpreted in the sense of an *equivalent control* as defined by Filippov; see Chapter 3.

More general *piecewise linear* characteristics can be modeled in a similar way. In this way, any dynamical input-state-output system with piecewise linear characteristics in the feedback loop can be represented as a hybrid system, with the locations corresponding to the different segments of the piecewise linear characteristics. For more information, especially with respect to well-posedness questions we refer to [33] and the references quoted there; see also Subsection 4.1.5.

2.2.12 Railroad crossing

Consider the railroad crossing from [2]. The system can be described as a conjunction of three subsystems, named 'train', 'gate', and 'controller'. The train sends a signal to the controller at least two minutes before it enters the crossing. Within one minute, the controller then sends a signal to the gate which is then closed within another minute. At most five minutes after it has announced its approach, the train has left the crossing and sends a corresponding signal to the controller. Within one minute the controller then provides a raise signal to the gate, which after receiving this signal takes one to two minutes to revert to the open position. Below a formal description is given in the form of an EFF in which expressions from the propositional calculus are freely used. Several variants are possible depending on the precise interpretation that one wants to give to the verbal description. The system is naturally described as a parallel composition of three subsystems:

$$\text{system} = \text{train} \parallel \text{gate} \parallel \text{controller} \tag{2.21a}$$

with

$$
\text{train} \left|
\begin{array}{l}
\dot{x} = 1, \quad \{Q_T = 1 \Rightarrow x \le 5\} \\
P_T = \text{out}, \quad S_T = \text{approach}, \quad x^+ = 0, \quad Q_T^+ = 1 \\
x \ge 2, \quad P_T^- = \text{out}, \quad P_T^+ = \text{in} \\
P_T^- = \text{in}, \quad P_T^+ = \text{out}, \quad S_T = \text{exit}, \quad Q_T^+ = 0
\end{array}
\right. \tag{2.21b}
$$

$$
\text{gate} \left|
\begin{array}{l}
\dot{y} = 1, \quad \{Q_{G1} = 1 \Rightarrow y \le 1\}, \quad \{Q_{G2} = 1 \Rightarrow y \le 2\} \\
S_C = \text{lower}, \quad y^+ = 0, \quad Q_{G1}^+ = 1 \\
P_G^- = \text{open}, \quad P_G^+ = \text{closed}, \quad Q_{G1}^+ = 0 \\
S_C = \text{raise}, \quad y^+ = 0, \quad Q_{G2}^+ = 1 \\
y \ge 1, \quad P_G^- = \text{closed}, \quad P_G^+ = \text{open}, \quad Q_{G2}^+ = 0
\end{array}
\right. \tag{2.21c}
$$

$$
\text{controller} \left|
\begin{array}{l}
\dot{z} = 1, \quad \{Q_C = 1 \Rightarrow z \le 1\} \\
S_T = \text{approach}, \quad z^+ = 0, \quad Q_C^+ = 1 \\
S_C = \text{lower}, \quad Q_C^+ = 0 \\
S_T = \text{exit}, \quad z^+ = 0, \quad Q_C^+ = 1 \\
S_C = \text{raise}, \quad Q_C^+ = 0.
\end{array}
\right. \tag{2.21d}
$$

The system above is nondeterministic. Note how enabling conditions are formulated in the event conditions, and enforcing conditions are placed in the flow conditions.

An important line of research consists in the development of software tools which can help in studying safety and liveness properties on the basis of formal descriptions like the one above for more complicated systems. Often such properties can be expressed as reachability properties; for instance, to use the examples given in [2], in the train-gate-controller system one may want to verify that the sets $\{P_T = \texttt{in},\ P_G = \texttt{open}\}$ and $\{P_G = \texttt{closed},\ y \geq 10\}$ are *not* reachable.

2.2.13   Power converter

Consider the power converter in Figure 2.9 (cf. [51]). The circuit in Figure 2.9

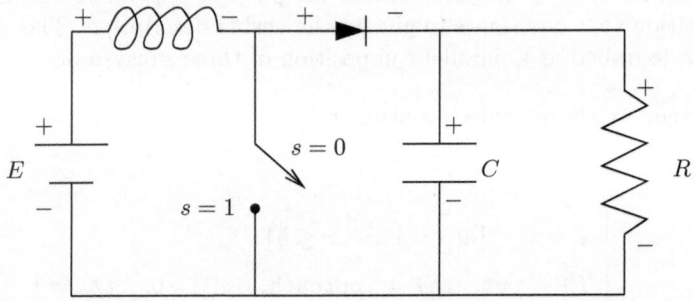

Figure 2.9: Boost circuit with clamping diode

consists of an inductor L with magnetic flux linkage ϕ_L, a capacitor C with electric charge q_C and a resistance load R, together with a diode and an ideal switch, with switch positions $s = 1$ (switch closed) and $s = 0$ (switch open). The diode is modeled as an ideal diode with voltage-current characteristic given by Figure 2.10. The constitutive relation of an ideal diode can be succinctly

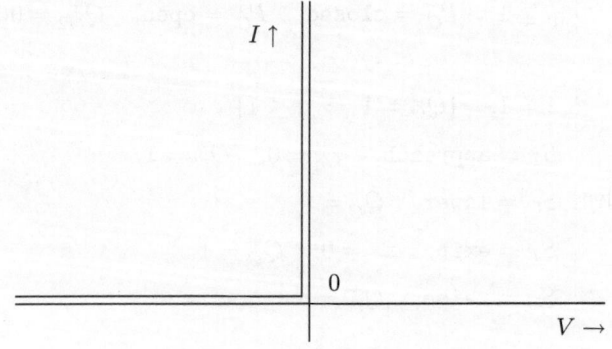

Figure 2.10: Ideal diode

expressed as follows:

$$v_D i_D = 0, \quad v_D \leq 0, \quad i_D \geq 0. \tag{2.22}$$

The circuit is used to obtain a voltage at the resistance load (the output voltage) that is higher than the voltage E of the input source; therefore it is commonly called a *step-up* converter.

Intuitively it is clear that the system can be represented as a hybrid system with four locations (or *modes*), corresponding to the two segments of the diode characteristic and the two switch positions. Furthermore, the transitions from a location with open switch to a location with closed switch, and vice versa, are *controlled* (externally induced), while the transitions corresponding to a change from one segment of the diode characteristic to another are *autonomous*.

Taking as continuous state (energy) variables the electric charge q_C and the magnetic flux ϕ_L, and as stored energy the quadratic function $\frac{1}{2C}q_C^2 + \frac{1}{2L}\phi_L^2$ we obtain the following dynamical equations of the circuit:

$$\begin{bmatrix} \dot{q}_C \\ \dot{\phi}_L \end{bmatrix} = \begin{bmatrix} -\frac{1}{R} & 1-s \\ s-1 & 0 \end{bmatrix} \begin{bmatrix} \frac{q_C}{C} \\ \frac{\phi_L}{L} \end{bmatrix} + \begin{bmatrix} 0 \\ 1 \end{bmatrix} E + \begin{bmatrix} si_D \\ (s-1)v_D \end{bmatrix}. \tag{2.23}$$

Here $s = 0,1$ denotes the switch, E is the voltage of the input source, and i_D, v_D, are respectively the current through the diode, and the voltage across the ideal diode. The dynamics of the circuit is completely specified by (2.23) together with the switch position (a discrete variable) and the constitutive relation of the ideal diode given by (2.22).

The separate dynamics of the four locations are obtained by substituting the following equalities into (2.23).

- Location 1 : $s = 0$, $v_D = 0$

- Location 2 : $s = 1$, $i_D = 0$

- Location 3 : $s = 0$, $i_D = 0$

- Location 4 : $s = 1$, $v_D = 0$

This yields for each of the four locations the following continuous dynamics:

$$\begin{aligned} \dot{q}_C &= -\tfrac{1}{L}\phi_L - \tfrac{1}{RC}q_C \\ \dot{\phi}_L &= -\tfrac{1}{C}q_C + E \end{aligned}$$

$$\begin{aligned} \dot{q}_C &= -\tfrac{1}{RC}q_C \\ \dot{\phi}_L &= E \end{aligned}$$

$$\begin{aligned} \dot{q}_C &= -\tfrac{1}{RC}q_C \\ \dot{\phi}_L &= 0 \end{aligned}$$

$$\dot{q}_C = 0$$
$$\dot{\phi}_L = E$$

In order to find the currently active location, we first observe the position of the switch. For $s = 0$ we have the locations 1 and 3. The location 1 is determined by $v_D = 0$ and $i_D = \frac{\phi_L}{L} \geq 0$. The latter inequality yields the location invariant $\phi_L \geq 0$. The location 3 is given by $i_D = \frac{\phi_L}{L} = 0$ and $v_D = E - \frac{q_C}{C} \leq 0$. This yields the location invariants $\phi_L = 0$ and $q_C - E \geq 0$. Similarly, if $s = 1$ then the system is in location 2 for $i_D = 0$ and the voltage $v_D = -\frac{q_C}{C} \leq 0$, giving the location invariant $q_C \geq 0$, and in location 4 if $v_D = \frac{q_C}{C} = 0$ and $i_D \geq 0$, leading to the location invariant $q_C = 0$.

Furthermore, it is straightforward (but tedious!) to write down all the transition guards, and the jump relations. In fact, it can be seen that in "normal operation", that means, if we start from an initial continuous state with $\phi_L \geq 0$ and $q_C \geq 0$ and — very importantly — with the input voltage $E \geq 0$, then no jumps will occur and $\phi_L(t) \geq 0$ and $q_C(t) \geq 0$ for all $t > 0$.

Let us note that one of the two location invariants of location 3, namely $q_C - E \geq 0$ explicitly depends on the external continuous variable E. Thus the example fits into the generalized hybrid automaton model, but not in the hybrid automaton model of Definition 1.2.3. Also, note that it makes sense to define the continuous state space of locations 3 and 4 to be given by the one-dimensional spaces $\{(q_C, \phi_L) \,|\, \phi_L = 0\}$, respectively $\{(q_C, \phi_L) \,|\, q_C = 0\}$, in accordance with the remark made before that in some cases it is natural to allow in the (generalized) hybrid automaton model for *different* state spaces for the various locations.

Equations (2.23) and (2.22) provide an (incomplete) event-flow formula description of the system; incomplete because the event-clauses corresponding to the diode have not been specified. The example demonstrates that the (generalized) hybrid automaton model may be far from an efficient description of a hybrid system: while the event-flow model given by (2.23) and (2.22) follows immediately from modeling, the hybrid automaton representation of this simple example already becomes involved.

2.2.14   Constrained pendulum

This example, which has been extensively discussed in [27], is used here to illustrate that the choice of state space variables may be crucial for the complexity of the resulting hybrid system description.[1] Consider a mathematical pendulum with length l that hits a pin such that the constrained pendulum has length l_c, cf. Figure 2.11. Taking as continuous state space variables $x = (\phi, v)$, where v is the angular velocity of the end of the pendulum, we

---

[1]This point was kindly brought to our attention by P.C. Breedveld.

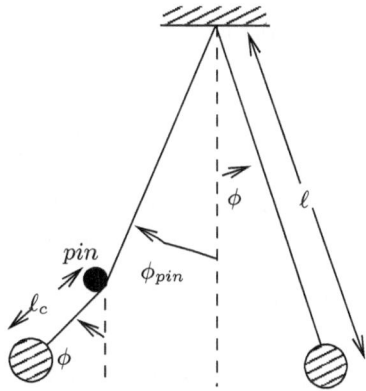

Figure 2.11: Pendulum constrained by a pin

obtain a hybrid system with two locations (unconstrained and constrained). The unconstrained dynamics, valid for $\phi \geq \phi_{pin}$, is given by

$$\begin{aligned} \dot{\phi} &= \tfrac{1}{l}v \\ \dot{v} &= -g\sin\phi - \alpha v \end{aligned}$$

(2.24)

for some friction coefficient α. The constrained dynamics, valid for $\phi \leq \phi_{pin}$, is given by the *same* equations with l replaced by l_c. It is immediately seen that there are no jumps in the continuous state vector x at the event times. Instead there is only a discontinuity in the righthand side of the differential equation caused by the change from l to l_c, or vice versa.

On the other hand, if we would take as continuous state space variables ϕ and $\dot{\phi}$, then at the moment that the swinging pendulum hits the pin, there is a jump in the second state space variable from $\dot{\phi}$ to $\frac{l}{l_c}\dot{\phi}$, and conversely, if the pendulum swings back from the constrained movement to the unconstrained movement there is a jump from $\dot{\phi}$ to $\frac{l_c}{l}\dot{\phi}$.

Clearly the resulting hybrid description is more complex from the hybrid systems point of view than the (equivalent!) description given before. On the other hand, from a physical modeling point of view the occurrence of jumps due to the "collision" of the rope with the pin is rather natural, while the "smart" choice of the continuous state variables in the first description eliminates in some way these jumps.

2.2.15 Degenerate Van der Pol oscillator

In this example (taken from [136]) we indicate that systems described by differential-algebraic equations may exhibit jump features, which could motivate a description as a hybrid system. Consider a degenerate form of the van der Pol oscillator consisting of a 1-F capacitor in parallel with a nonlinear resistor with a specific characteristic:

$$\begin{aligned}
\dot{v} &= i \\
v &= -\tfrac{1}{3}i^3 + i.
\end{aligned}$$

(2.25)

These equations are interpreted as an implicitly defined dynamics on the one-dimensional constraint submanifold C in (v, i) space given by

$$C = \{(v, i) \,|\, v = -\frac{1}{3}i^3 + i\}.$$

Difficulties in this interpretation arise in the points $(-1, -\frac{2}{3})$ and $(1, \frac{2}{3})$. At these points $\dot{v}$ is negative, respectively positive, while the corresponding time-derivative of i in these points is positive, respectively negative. Hence, because of the form of the constraint manifold C it is not possible to "integrate" the dynamics from these points in a continuous manner along C.

Instead it has been suggested in [136] (see this paper for other related references) that a suitable interpretation of the dynamics from these singular points is given by the following jump rules:

$$\begin{aligned}
(-1, -\tfrac{2}{3}) &\rightarrow (2, -\tfrac{2}{3}) \\
(1, \tfrac{2}{3}) &\rightarrow (-2, \tfrac{2}{3})
\end{aligned}$$

(2.26)

Alternatively, the resulting system can be described as a hybrid system with two locations with continuous state spaces both given by C and dynamics described by (2.25), with location invariants $i \leq -1$, respectively $i \geq 1$, and jump relations given by (2.26).

2.3   Notes and References for Chapter 2

The choice of the examples in this chapter clearly reflects the interests and bias of the authors of this text. A wealth of different examples of hybrid systems can be found in the literature, see in particular the proceedings [58], [5], [3], [100], [6], [137], [153], as well as the journal special issues [7] and [116]. For further examples and references in the context of mechanical problems, an excellent source is [31].

Chapter 3

Variable-structure systems

Discontinuous dynamical systems have been an object of systematic study in the former Soviet Union and other Eastern European countries for a long period starting in the late 1940s. Much of the theory has been developed in close connection with control theory, which has provided many motivating examples such as relay controls and bang-bang controls in shortest-time problems. Systematic expositions of the results of this research are available in a number of textbooks, for instance the ones by Andronov et al. [4], Tsypkin [150], Utkin [152], and Filippov [54]. Here we shall not attempt to summarize all this work; instead we concentrate on the solution concept for discontinuous dynamical systems, and more specifically on the so-called "sliding mode". We follow mainly Filippov's book. We do not aim for the greatest possible generality; in particular we limit ourselves to systems with constant parameters.

3.1   Discontinuous dynamical systems

A typical example of the type of systems considered by Filippov can be constructed as follows. Let S_0 be a surface in n-dimensional space, by which we mean that S_0 is a $(n-1)$-dimensional differentiable manifold determined as the null set of a smooth real-valued function ϕ on $\mathbb{R}^n$. The set of all x for which $\phi(x)$ is positive (negative) will be denoted by S_+ (S_-). Let f_+ be a continuous function defined on $S_+ \cup S_0$, let f_- be a continuous function defined on $S_- \cup S_0$, and let f be the function defined on $S_+ \cup S_-$ by $f(x) = f_+(x)$ for $x \in S_+$ and $f(x) = f_-(x)$ for $x \in S_-$. It is not required that the functions f_+ and f_- agree on S_0, so that in general the function f cannot be extended to a continuous function defined on all of $\mathbb{R}^n$. Now consider the differential equation

$$\dot{x}(t) = f(x(t)). \tag{3.1}$$

Here we have a dynamic system whose dynamics changes abruptly when the state vector crosses the switching surface S_0. In a control context, such a situation could occur as a result of a gain scheduling control law which switches from one feedback to another when a certain function of the state variables crosses a certain threshold. In general it cannot be expected that state trajectories are differentiable at points where the boundary is crossed, and so it would be too much to require the validity of (3.1) for all t. One possible way out is to replace the equation (3.1) by the integral form

$$x(t) \; = \; x(0) + \int_0^t f(x(s))ds \tag{3.2}$$

which doesn't require the trajectory $x(\cdot)$ to be differentiable. When differential equations are interpreted in this way they are sometimes called *Carathéodory equations*. The interpretation (3.2) also obviates the need for specifying the value of f on the surface S_0, at least for cases in which solutions arrive at this surface from one side and leave it immediately on the other side.

Whether or not we do have solutions that cross the switching surface instantaneously depends on the vector fields determined by f_+ and f_-. Consider a point x_0 on the switching surface S_0. At this point we have two vectors $f_+(x_0)$ and $f_-(x_0)$. In terms of these vectors and their relation to the tangent space of the surface S_0 at the point x_0 we can distinguish the following four main cases.

(i) Both vectors point inside S_+. In this case, state trajectories can only arrive at x_0 from S_-, and will continue in S_+. The Carathéodory interpretation is sufficient.

(ii) Both vectors point inside S_-; this is analogous to case (i).

(iii) The vector $f_+(x_0)$ points inside S_+ and the vector $f_-(x_0)$ points inside S_-. In this case x_0 cannot be reached by trajectories of (3.1). If x_0 is taken as an initial condition, there are two possible solutions to (3.2).

(iv) The vector $f_+(x_0)$ points inside S_- and the vector $f_-(x_0)$ points inside S_+. In this case the Carathéodory interpretation does not give us a usable solution concept.

Filippov is mainly concerned with situations in which case (iv) occurs. It can be argued that it is physically meaningful to try to develop also a solution concept for this case. For instance, it may be that the function f is a simplified version of another function $\tilde{f}$ that is not actually discontinuous across the switching surface, but that changes in a steep gradient from f_+ on one side to f_- on the other side of S_0. The solutions of the differential equation $\dot{x} = \tilde{f}(x)$ will in case (iv) tend to follow the switching surface since they are "pushed" onto S_0 from both sides. In another interpretation, suppose that the transition from the regime described by f_+ to the one described by f_- is caused by a switching controller that monitors the sign of $\phi(x(t))$. For a number of practical reasons, switching will not occur exactly when $\phi(x(t))$ crosses the zero value, but at some nearby instant of time. In case (iv) the result will be a "chattering" behavior in which the system switches quickly from one dynamics to the other and back again. Also in this way a motion will result which will take place more or less along the switching surface. So on the one hand there are good reasons to allow for solutions along the switching surface in cases of type (iv), on the other hand several different physical mechanisms may be at work, and one can also say that the situation is perhaps not completely specified by the two functions f_+ on S_+ and f_- on S_-. Therefore Filippov actually discusses several different solution concepts.

3.2 Solution concepts

Let us first look at what Filippov calls the *simplest convex definition*. Take the situation of case (iv); so consider a point x_0 on the switching surface, and suppose that the vectors $f_+(x_0)$ and $f_-(x_0)$ point inside S_- and S_+ respectively. Since these two vectors are at different sides of the tangent space of S_0 at x_0, there must be a convex combination of them which lies in the tangent space. Denote the vector obtained in this way by $f_0(x_0)$. Repeating the construction for points x on S_0 in a neighborhood of x_0, we obtain a function $f_0(x)$ defined on S_0 at least in a neighborhood of x_0, and having the property that the vector $f_0(x)$ always points in the tangent space of S_0 at x. Therefore, the differential equation $\dot{x} = f_0(x)$ can be used to define a motion on S_0 which is called a *sliding motion*. The concept of solution is now extended to include this type of motion as well.

The second notion of solution discussed by Filippov uses the so-called *equivalent control method*. For the application of this method, the function f is supposed to be of the form $f(x, u(x))$ where $u(\cdot)$ is a multivalued function that is in fact single-valued on $S_+ \cup S_-$ but that has a range of values $U(x)$ for $x \in S_0$. The set $U(x)$ is a closed interval. In the situation of case (iv) above, an "equivalent control" $u_{eq}(x)$ is sought for $x \in S_0$ such that $f(x, u_{eq}(x))$ is tangent to S_0 and $u_{eq}(x) \in U(x)$. The motion given by $\dot{x} = f(x, u_{eq}(x))$, which is a motion along the sliding surface as in the case of the simplest convex definition, is then used to define a notion of solution. The resulting motion may be quite different from the one produced by the first definition; see Examples 3.2.1 and 3.2.2 below.

Filippov also considers a third definition. This definition again starts from the formulation $\dot{x} = f(x, u(x))$ with $u(x) \in U(x)$, where $U(x)$ is a single point for $x \in S_+ \cup S_-$ and is a closed interval for $x \in S_0$. For given x_0, let $F(x_0)$ denote the smallest convex set containing $\{f(x_0, u) \mid u \in U(x_0)\}$. We can now consider the differential inclusion $\dot{x}(t) \in F(x(t))$. Away from the switching surface, this inclusion is in fact a standard differential equation since $F(x)$ then consists of only a single point. On the switching surface, the requirement $\dot{x}(t) \in F(x(t))$ leaves considerable latitude; however, solution trajectories must still follow the switching manifold in the neighborhood of points where case (iv) applies, because solutions that enter either S_+ or S_- are not possible.

In case $f(x, u)$ depends affinely on u and the interval $U(x_0)$ is $[u_+, u_-]$ with $u_+ = \lim_{x \in S_+, x \to x_0} u(x)$ and u_- defined likewise, all solution concepts are the same. In other cases however, the third definition does not uniquely determine the velocity of the motion along the switching surface. The indeterminism that is introduced this way may be viewed as a way of avoiding the choice between the other two solution concepts. Such a conservative stance can be well-motivated in a verification analysis in which one would like to build in a degree of robustness against model misspecification. From a point of view of simulation however, one would rather work with uniquely defined trajectories.

It might be taken as an objection against the equivalent control method that it requires the modeler to specify a function $f(x, u)$ and a function $u(x)$ on $S_+ \cup S_-$ together with a closed interval $U(x)$ for $x \in S_0$; in this way, the modeler is forced into decisions that he or she would perhaps prefer to avoid. The simple convex definition only requires the specification of functions f_+ on $S_+ \cup S_0$ and f_- on $S_- \cup S_0$. It may be argued however that assuming the simple definition comes down to making a particular choice for the functions $f(x, u)$ and $u(x)$ that are used in the equivalent control method. For this purpose, assume that the continuous functions f_+ and f_- defined on $S_+ \cup S_0$ and $S_- \cup S_0$ respectively are extended in some arbitrary way to continuous functions to all of $\mathbb{R}^n$. If we now define $f(x, u)$ by

$$f(x, u) = \tfrac{1}{2}(1 + u)f_+(x) + \tfrac{1}{2}(1 - u)f_-(x) \tag{3.3}$$

then f is continuous as a function of x and u. Furthermore define $u(x) = \operatorname{sgn}(\phi(x))$, where sgn is the multivalued function defined by (1.13). Note that $f(x, u)$ is affine in x. It is easily verified that the equivalent control method and the third definition now both generate solutions that coincide with the ones obtained from the simple convex definition.

To illustrate the difference between the equivalent control method and the simplest convex definition, we present the following two examples.

Example 3.2.1. This example is taken from [109]. Let a system be given by

$$\dot{x}_1 = \cos \theta u, \quad \dot{x}_2 = -\sin \theta u, \quad y = x_2, \quad u = \operatorname{sgn} y. \tag{3.4}$$

Both for $x_2 > 0$ and for $x_2 < 0$ the right hand side is constant and so the system above could be called a "piecewise constant system". The trajectories are straight lines impinging on the x_1-axis at an angle determined by the parameter θ which is chosen from $(0, \pi)$. Along the x_1-axis, a sliding mode is possible. The equivalent control method applied to (3.4) determines u such that $\dot{x}_2 = -\sin \theta u = 0$; obviously this requires $u = 0$ so that the sliding mode is given by $\dot{x}_1 = 1$. If one would take the simplest convex definition, one would look for a convex combination of the vectors $\operatorname{col}(\cos \theta, -\sin \theta)$ and $\operatorname{col}(\cos \theta, \sin \theta)$ such that the second component vanishes. There is one such convex combination, namely

$$\operatorname{col}(\cos \theta, 0) = \tfrac{1}{2}\operatorname{col}(\cos \theta, -\sin \theta) + \tfrac{1}{2}\operatorname{col}(\cos \theta, \sin \theta).$$

In this case the sliding mode is given by $\dot{x}_1 = \cos \theta$. The question which of the two sliding modes is the "correct" one has no general answer; different approximations of the relay characteristic may lead to different sliding modes. The third definition leads to the differential inclusion $\dot{x}_1 \in [\cos \theta, 1]$.

Let us now consider a smooth approximation as well as a chattering approximation to the sliding mode. In the smooth approximation, we assume that there is a very quick but continuous transition from the vector field on one side of the switching surface to the vector field on the other side. This may be effectuated by replacing the relation $u = \operatorname{sgn} y$ by a sigmoid-type function, for instance

$$u = \tanh(y/\varepsilon) \tag{3.5}$$

where ε is a small positive number. Doing this for (3.4) we get the smooth dynamical system

$$\dot{x}_1 = \cos(\theta \tanh x_2), \quad \dot{x}_2 = -\sin(\theta \tanh x_2) \tag{3.6}$$

whose trajectories are very similar to those of (3.4) at least outside a small band along the switching curve $x_2 = 0$. The system (3.6) has solutions $x_1(t) = t + c$, $x_2(t) = 0$ which satisfy the equations of the sliding mode according to the equivalent control definition.

Consider next the chattering approximation. Again we choose a small positive number ε, and we define a "chattering system" by the event-flow formula

$$\begin{vmatrix} P = \text{up}, & \dot{x}_1 = \cos\theta, & \dot{x}_2 = -\sin\theta, & x_2 \geq -\varepsilon \\ P = \text{down}, & \dot{x}_1 = \cos\theta, & \dot{x}_2 = \sin\theta, & x_2 \leq \varepsilon \\ x_2 = -\varepsilon, & P^- = \text{up}, & P^+ = \text{down} \\ x_2 = \varepsilon, & P^- = \text{down}, & P^+ = \text{up.} \end{vmatrix} \tag{3.7}$$

The trajectories of this system are exactly the same as those of the original system (3.4) except in a band of width 2ε around the switching curve. For small ε the system (3.7) has solutions that are close to trajectories of the form $x_1(t) = t \cos\theta + c$, $x_2(t) = 0$. These trajectories satisfy the equations of the sliding mode according to the simplest convex definition.

It may be noted that (3.6) is not the only possible smooth approximation to (3.4); another possibility is for instance

$$\dot{x}_1 = \cos\theta, \quad \dot{x}_2 = -\tanh(x_2/\varepsilon)\sin\theta. \tag{3.8}$$

The trajectories of this system are arbitrarily close to those of (3.4) outside a band around the switching curve if ε is taken small enough. The system (3.8) has solutions of the form $x_1(t) = t \cos\theta + c$, $x_2(t) = 0$ which conform to the equations of the sliding mode according to the simplest convex definition.

The solution according to the simplest convex definition can be obtained as an equivalent control solution if we replace the equations (3.4) by

$$\dot{x}_1 = \cos\theta, \quad \dot{x}_2 = -u\sin\theta, \quad y = x_2, \quad u = \operatorname{sgn} y. \tag{3.9}$$

Actually in this case the smooth approximation according to the recipe (3.5) leads to the system (3.8).

Example 3.2.2. Consider the system

$$\dot{x}_1(t) = -x_1(t) + x_2(t) - u(t) \tag{3.10}$$

$$\dot{x}_2(t) = 2x_2(t)(u^2(t) - u(t) - 1) \tag{3.11}$$

$$u(t) = \operatorname{sgn} x_1(t). \tag{3.12}$$

The system has a sliding mode on the interval $x_1 = 0$, $-1 \le x_2 \le 1$. According to the simplest convex definition, the sliding mode is given by

$$\dot{x}_2 = -2x_2^2 \tag{3.13}$$

wheras according to the equivalent control method the sliding mode is given by

$$\dot{x}_2 = 2x_2(x_2^2 - x_2 - 1). \tag{3.14}$$

The two dynamics (3.13) and (3.14) are quite different; (3.13) has an unstable equilibrium at 0 whereas (3.14) has two equilibria, of which the one at $\frac{1}{2} - \frac{1}{2}\sqrt{5}$ is unstable and the one at 0 is stable. In particular, solutions of the system (3.10) in the "simplest convex" interpretation and in the "equivalent control" interpretation that are identical until they reach a point on the x_2-axis with $\frac{1}{2} - \frac{1}{2}\sqrt{5} \le x_2 \le 0$ will after this point follow entirely different paths. The equivalent control solution will be recovered by a smoothing approximation such as $u = \tanh(x_1/\varepsilon)$, whereas other methods that are based on some type of sampling will follow the solution according to the simplest convex definition.

3.3   Reformulations

The generality in the formulation of the equivalent control method may be reduced somewhat without loss of expressive power. As above, let u_+ and u_- be extended arbitrarily to continuous functions on all of $\mathbb{R}^n$. Introduce a new variable v, and define a continuous function $\tilde{f}$ of the two variables x and v by

$$\tilde{f}(x,v) = f(x, \tfrac{1}{2}(1-v)u_-(x) + \tfrac{1}{2}(1+v)u_+(x)).$$

Given that f is continuous as a function of x and u, the function $\tilde{f}$ will be continuous as a function of x and v. Moreover, we have $\tilde{f}(x,1) = f(x, u_+(x))$ and $\tilde{f}(x,-1) = f(x, u_-(x))$. In the most important special case in which the end points of the interval $U(x)$ for $x \in S_0$ are $u_+(x)$ and $u_-(x)$, we can simply define $v(x) = \operatorname{sgn}(\phi(x))$ to get the same solutions to $\dot{x} = \tilde{f}(x, v(x))$ as one would get from $\dot{x} = f(x, u(x))$ according to the equivalent control method.

In Filippov's treatment, the word *mode* is used (in the term "sliding mode") but discontinuous systems are not modeled explicitly as hybrid systems. By rewriting the equations in terms of event-flow formulas, more emphasis is placed on the multimodal aspects. In the equivalent control method, we are given a function $f(x,u)$ and a multivalued function $u(x)$. On S_+ and

S_-, u is single-valued; denote the corresponding functions on S_+ and S_- by u_+ and u_- respectively. For $x \in S_0$, u can take values in a closed interval $U(x) =: [\bar{u}_+(x), \bar{u}_-(x)]$. It seems reasonable to conjecture (but of course it needs proof) that, under fairly general circumstances, Filippov's solutions according to the equivalent control definition correspond to the continuous state traces of the solutions in $Z/1/C^{1/0}/C^0$ of the EFF

$$
\left\| \begin{array}{ll}
\dot{x} = f(x, u), \quad y = \phi(x) \\[4pt]
\left\| \begin{array}{ll}
y > 0, \quad u = u_+(x) \\[2pt]
y < 0, \quad u = u_-(x) \\[2pt]
y = 0, \quad \bar{u}_-(x) \leq u \leq \bar{u}_+(x)
\end{array} \right.
\end{array} \right. \tag{3.15}
$$

For the solution concept corresponding to the simple convex definition one would rather use the following EFF:

$$
\left\| \begin{array}{ll}
\dot{x} = \frac{1}{2}(1 + u)f_+(x) + \frac{1}{2}(1 - u)f_-(x), \quad y = \phi(x) \\[4pt]
\left\| \begin{array}{ll}
y > 0, \quad u = 1 \\[2pt]
y < 0, \quad u = -1 \\[2pt]
y = 0, \quad -1 \leq u \leq 1
\end{array} \right.
\end{array} \right. \tag{3.16}
$$

The three-term disjunction in the parts corresponding to non-event times indicates a system with three modes, one corresponding to motion inside S_+, another corresponding to motion inside S_-, and a sliding mode. In the sliding mode, trajectories must move along the switching manifold characterized by $y = 0$, and this motion must take place along a vector of the form $f(x, u)$ with $u \in U(x) = [\bar{u}_-(x), \bar{u}_+(x)]$ or along a suitable convex combination of vectors $f_+(x)$ and $f_-(x)$. The motion on S_+ and S_- follows the differential equations $\dot{x} = f_+(x)$ and $\dot{x} = f_-(x)$ respectively. All this is expressed in (3.15) and (3.16).

It should be emphasized that, although the definitions have been designed to avoid some obvious cases of nonexistence of solutions, neither existence nor uniqueness of solutions to the type of systems considered in this section is automatic. We shall come back to this issue below.

3.4 Systems with many regimes

So far we have been discussing situations in which there is one surface that determines the various regimes under which the system can evolve. Such a situation occurs for instance when friction at one point in a mechanical system is modeled by the Coulomb friction law. Of course we can easily have situations in which there are many points of friction and then the space through which

the continuous state variables of the system move will be divided in many
parts each with their own dynamical regime. Suppose that we have k surfaces
determined by functions $\phi_1, \ldots, \phi_k$ from $\mathbb{R}^n$ to $\mathbb{R}$. In principle, one should now
allow sliding modes not just along each of the surfaces $S_0^i := \{x \mid \phi_i(x) = 0\}$
but also along the intersection $S_0^i \cap S_0^j$ of two surfaces, which will in general
be a manifold of dimension $n - 2$, as well as along the intersection of three
surfaces and so on. To create these sliding modes, one will in general need k
independent control variables, so the given dynamics should be of the form

$$\dot{x} = f(x, u_1, \ldots, u_r)$$

with $r \geq k$; moreover, there should be enough freedom in the variables $u_i(x)$
when x is in the intersection of several surfaces S_0^i to allow motion along
these intersections. If one wants to ensure uniqueness of solutions, then it
is natural to let r be equal to k and also to establish a one-one connection
between control variables and surfaces. The existence of such a connection,
which ties each input variable u_i to a corresponding "output variable" defined
by $y_i = \phi_i(x)$, is anyway natural in applications such as relay systems and
Coulomb friction. An event-flow formula may be written down as follows:

$$\dot{x} = f(x, u) \|_{i \in \overline{k}} \{ y_i = \phi_i(x), \ (y_i > 0, \ u_i = u_{i+}(x)) \mid$$
$$(y_i < 0, \ u_i = u_{i-}(x)) \mid (y_i = 0, \ \overline{u}_{i-}(x) \leq u_i \leq \overline{u}_{i+}(x)) \} \quad (3.17)$$

where $\overline{k}$ denotes the set $\{1, \ldots, k\}$; solutions are sought in a space with continuous state trajectories. As above, in case $\overline{u}_{i+}(x) = u_{i+}(x)$ and $\overline{u}_{i-}(x) = u_{i-}(x)$
for all i and all $x \in S_0$, it is possible to rewrite the system in such a way that
the control variables are related to the output variables $\phi_i(x)$ by the signum
function sgn. The discontinuous system is then rewritten as a relay system.

In the case of several control variables the relation between the equivalent control method and the simplest convex definition can be more complicated than in the case of a single control variable. Consider the example
$\dot{x} = f(x, u_1, u_2)$ with $u_i = \text{sgn}(\phi_i(x))$ for $i = 1, 2$. If $\phi_1(x) = \phi_2(x)$ the
state space is actually divided into two parts, and the dynamics is given by
$\dot{x} = f(x, 1, 1)$ on one side and by $\dot{x} = f(x, -1, -1)$ on the other side of the
dividing surface. The simple convex definition would require a sliding motion
along the surface to be generated by some convex combination of the two vectors $f(x, 1, 1)$ and $f(x, -1, -1)$, whereas the equivalent control method would
allow the motion to be generated by an arbitrary vector of the form $f(x, u_1, u_2)$
with $-1 \leq u_1 \leq 1$ and $-1 \leq u_2 \leq 1$. Even in the case in which f depends
linearly on u_1 and u_2 the allowed motions according to the two definitions are
in general different. The example is perhaps artificial though; in fact with the
simple convex definition there is little reason to consider the system as one
having two control variables. In case the system is interpreted as having two
control variables and the equivalent control method is used to define solutions,
it is likely that the solutions are not uniquely determined since in the sliding
mode there are two control variables available to satisfy only one constraint.

3.5   The Vidale-Wolfe advertizing model

As an example of a situation in which multimodality arises in a natural way, let us consider the following optimization problem which was suggested in 1957 by Vidale and Wolfe as a simple model for supporting decisions on marketing expenditures. The problem is:

$$\text{maximize} \qquad \int_0^T (x(t) - cu(t))dt \tag{3.18a}$$

$$\text{subject to} \qquad \dot{x}(t) = -ax(t) + (1 - x(t))u(t) \tag{3.18b}$$

$$x(0) = x_0 \in (0, 1) \tag{3.18c}$$

$$0 \le u(t) \le 1. \tag{3.18d}$$

The state variable $x(t)$ represents market share, whereas the control variable $u(t)$ represents marketing effort, normalized in such a way that saturation occurs at the value 1. The constant c expresses cost of advertizing, and a indicates the rate at which sales will decrease when no marketing is done. A standard application of the maximum principle (see for instance [142]) suggests the following procedure for finding candidate optimal solutions. First define the Hamiltonian

$$H(x, u, \lambda) = x - cu + \lambda(-ax + (1 - x)u) \tag{3.19}$$

where $\lambda(\cdot)$ is an adjoint variable. According to the maximum principle, necessary conditions for optimality are given (in the "normal" case) by the equations (in shorthand notation)

$$\dot{x} = \frac{\partial H}{\partial \lambda}, \quad \dot{\lambda} = -\frac{\partial H}{\partial x}, \quad x(0) = x_0, \quad \lambda(T) = 0, \quad u = \arg \max_u H. \tag{3.20}$$

Since the Hamiltonian is linear in the control variable u, maximization over u leads to a relay-like characteristic:

$$\begin{vmatrix} u = 0, & -c + \lambda(1 - x) < 0 \\ u = 1, & -c + \lambda(1 - x) > 0 \\ 0 \le u \le 1, & -c + \lambda(1 - x) = 0. \end{vmatrix} \tag{3.21}$$

It will be convenient to introduce a function C by

$$C(x, \lambda) := -c + \lambda(1 - x). \tag{3.22}$$

The relay characteristic is connected to the differential equations

$$\dot{x} = -ax + (1 - x)u \tag{3.23}$$

$$\dot{\lambda} = -1 + (a + u)\lambda \tag{3.24}$$

which we may also write in vector form

$$\frac{d}{dt} \begin{bmatrix} x \\ \lambda \end{bmatrix} = \begin{bmatrix} -ax \\ a\lambda - 1 \end{bmatrix} + \begin{bmatrix} 1 - x \\ \lambda \end{bmatrix} u. \tag{3.25}$$

In the phase plane we clearly have a switching curve given by the equation $C(x, \lambda) = 0$. In the region defined by $(1 - x)\lambda < c$ we have the dynamics

$$\dot{x} = -ax, \quad \dot{\lambda} = a\lambda - 1 \tag{3.26}$$

whereas in the region $(1 - x)\lambda > c$ we have

$$\dot{x} = -(a + 1)x + 1, \quad \dot{\lambda} = (a + 1)\lambda - 1. \tag{3.27}$$

We have to determine the relation of the two vector fields that are defined in this way to the switching curve. For this purpose we compute the time derivative of the function $y(t) := C(x(t), \lambda(t))$ along the trajectories of each of the two dynamics given by (3.26) and (3.27). Actually it turns out that the result is the same in both cases and in fact doesn't even depend on u, since the gradient of C which is given by

$$[\frac{\partial C}{\partial x} \quad \frac{\partial C}{\partial \lambda}] = [-\lambda \quad 1 - x] \tag{3.28}$$

is orthogonal to the input vector field in (3.25). We get

$$\dot{y} = a\lambda + x - 1. \tag{3.29}$$

So the mode selected at a given point on the switching curve is determined by the sign of the quantity $a\lambda + x - 1$. If $a\lambda + x - 1 > 0$, then we enter the region $(1 - x)\lambda > c$ corresponding to the mode $u = 1$. If $a\lambda + x - 1 < 0$ then the selected mode is the one in which $u = 0$. Since both dynamics agree on the sign of $\dot{y}$ the sliding mode cannot occur as long as $a\lambda + x - 1 \neq 0$.

We still have to consider the possibility of a sliding mode at the point on the switching curve at which $a\lambda + x - 1 = 0$, if such a point does exist. It is easily verified that there is indeed such a point in the region of interest $(0 \leq x \leq 1)$ if $ac \leq 1$. Because we already know that a sliding mode cannot exist at other points in the (x, λ)-plane, a sliding regime must create an equilibrium. Since at the intersection point of the curves $y = 0$ and $\dot{y} = 0$ the gradient of C annihilates not only the input vector in (3.25) but also the drift vector, these two vectors are linearly dependent. This means that the two tangent vectors corresponding to $u = 0$ and $u = 1$ are also linearly dependent. If these vectors point in opposite directions, we can indeed create a sliding mode by choosing $u \in [0, 1]$ such that the right hand side in (3.25) vanishes. The three equations in three unknowns given by

$$\begin{aligned}
-ax + (1 - x)u &= 0 \\
a\lambda - 1 + \lambda u &= 0 \\
\lambda(1 - x) &= c
\end{aligned}$$

are solved in the region $0 \leq x \leq 1$ by

$$x = 1 - \sqrt{ac}, \quad \lambda = \sqrt{c/a}, \quad u = \sqrt{a/c} - a. \tag{3.30}$$

So the sliding mode exists when $0 \leq \sqrt{a/c} - a \leq 1$, that is, when

$$\frac{a}{(a+1)^2} \leq c \leq \frac{1}{a}. \tag{3.31}$$

Assume now that this condition holds, and consider the dynamics (3.21–3.25) with the initial condition

$$x_0 = 1 - \sqrt{ac}, \quad \lambda_0 = \sqrt{c/a}. \tag{3.32}$$

By computing $\ddot{y}$ it can be verified that from this initial conditions there are three possible continuations, corresponding to the three choices $u = 0$, $u = 1$ and $u = \sqrt{a/c} - a$. This actually means that there are infinitely many solutions that have initial condition (3.32), since one can take $u(t) = \sqrt{a/c} - a$ for some arbitrary positive time and then continue with either $u = 0$ or $u = 1$. Some of the trajectories of the system (3.21–3.25) for the parameter values $a = \frac{1}{2}$ and $c = 1$ are illustrated in Fig. 3.1.

We conclude that the system (3.21–3.25) is *not* well-posed as an initial-value problem if the parameters a and c satisfy (3.31). However it should be noted that the necessary conditions for the original optimization problem do not have the form of an initial-value problem but rather of a mixed boundary value problem, since we have the mixed boundary conditions $x(0) = x_0$ and $\lambda(T) = 0$. As such the problem *is* well-posed and the option of "lying still" at the equilibrium point created by the sliding mode is crucial. Also, the meaning of this equilibrium point and the corresponding value of u is clear: apart from "initial" and "final" effects, this value of u is the one that indicates optimal advertizing expenditures. Note that it would be easy to rewrite the suggested optimal policy as a feedback policy, with a feedback mapping that depends discontinuously on x and t, and that assumes only three different values.

3.6 Notes and references for Chapter 3

This chapter obviously leans heavily on the work by Filippov as described in his book [54]. The book, which appeared first in Russian in 1985, fits into a tradition of research in non-smooth dynamical systems that spans several decades. The solution concept proposed by Filippov originally in 1960 [53] uses the theory of differential inclusions, which is already in itself an important branch of mathematical analysis; see for instance [12]. In the present book our point of view is somewhat different. Rather than trying to infer in some way the behavior of the sliding mode which takes place on a lower-dimensional manifold from the system's behavior in the "main" modes which take place on full-dimensional regions, we consider all modes in principle on the same

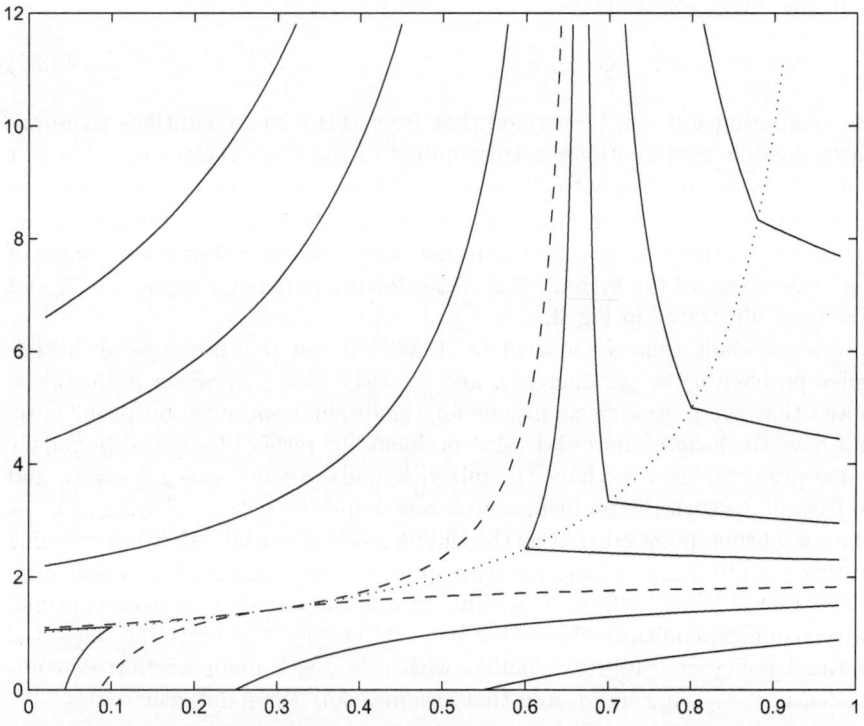

Figure 3.1: Trajectories for Vidale-Wolfe example. Horizontal: x, vertical: λ. Dotted: switching curve, with $u = 0$ to the right and $u = 1$ to the left. Dashed: trajectories leading to and from the equilibrium point

footing. We believe that this "hybrid" perspective is natural in many cases and can also be used effectively for the type of systems considered by Filippov and other authors in the same tradition. The hybrid point of view will be worked out further in the following chapters.

Chapter 4

Complementarity systems

We have already seen several examples of situations in which modes are determined by pairs of so-called "complementary variables". Two scalar variables are said to be *complementary* if they are both subject to an inequality constraint, and if at all times at least one of these constraints is satisfied with equality. The most obvious example is that of the ideal diode. In this case the complementary variables are the voltage across the diode and the current through it. When the voltage drop across the diode is negative the current must be zero, and the diode is said to be in *nonconducting mode*; when the current is positive the voltage must be zero, and the diode is in *conducting mode*. There are many more examples of hybrid systems in which mode switching is determined by complementarity conditions. We call such systems *complementarity systems*. As we shall see, complementarity conditions arise naturally in a number of applications; moreover, in several other applications one may rewrite a given system of equations and inequalities in complementarity form by a judicious choice of variables.

As a matter of convention, we shall always normalize complementary variables in such a way that both variables in the pair are constrained to be nonnegative; note that this deviates from standard sign conventions for diodes. So a pair of variables (u, y) is said to be subject to a complementarity condition if the following holds:

$$u \geq 0, \quad y \geq 0, \quad \left| \begin{array}{l} y = 0 \\ u = 0. \end{array} \right. \tag{4.1}$$

Often we will be working with several pairs of complementary variables. For such situations it is useful to have a vector notation available. We shall say that two vectors of variables (of equal length) are *complementary* if for all i the pair of variables (u_i, y_i) is subject to a complementarity condition. In the mathematical programming literature, the notation

$$0 \leq y \perp u \geq 0 \tag{4.2}$$

is often used to indicate that two vectors are complementary. Note that the inequalities are taken in a componentwise sense, and that the usual interpretation of the "perp" symbol (namely $\sum_i y_i u_i = 0$) does indeed, in conjunction

with the inequality constraints, lead to the condition $\{y_i = 0\} \vee \{u_i = 0\}$
for all i. Alternatively, one might say that the "perp" relation is also taken
componentwise.

Therefore, complementarity systems are systems whose flow conditions can
be written in the form

$$f(\dot{x}, x, y, u) = 0 \tag{4.3a}$$
$$0 \leq y \perp u \geq 0. \tag{4.3b}$$

In this formulation, the variables y_i and u_i play completely symmetric roles.
Often it is possible to choose the denotations y_i and u_i in such a way that the
conditions actually appear in the convenient "semi-explicit" form

$$\dot{x} = f(x, u) \tag{4.4a}$$
$$y = h(x, u) \tag{4.4b}$$
$$0 \leq y \perp u \geq 0. \tag{4.4c}$$

The flow conditions (4.3) or (4.4) still have to be supplemented by appropriate
event conditions which describe what happens when there is a switch between
modes. In some applications it will be enough to work with the default event
conditions that require continuity across events; in other applications one needs
more elaborate conditions.

Additional continuous input (or "control") variables may of course easily
be added to a system description such as (4.4); discrete input variables might
be added as well. In this chapter, however, we shall mainly be concerned with
closed systems in which such additional inputs do not appear. The motivation
for doing this is that we need an understanding of the dynamics of closed
systems before we can discuss systems with inputs. It may be noted, though,
that the dynamical system (4.3a) (or (4.4a–4.4b)) taken as such is an open
system, which is "closed" by adding the complementarity conditions (4.4c).
Therefore, the theory of open (or "input-output") dynamical systems will still
play an important role in this chapter.

In the mathematical programming literature, the so-called *linear complementarity problem* (LCP) has received much attention; see the book [39] for
an extensive survey. The LCP takes as data a real k-vector q and a real $k \times k$
matrix M, and asks whether it is possible to find k-vectors u and y such that

$$y = q + Mu, \quad 0 \leq y \perp u \geq 0. \tag{4.5}$$

The main result on the linear complementarity problem that will be used below
is the following [135], [39, Thm. 3.3.7]: the LCP (4.5) is uniquely solvable for
all data vectors q if and only if all principal minors of the matrix M are
positive. (Given a matrix M of size $k \times k$ and two nonempty subsets I and
J of $\{1, \ldots, k\}$ of equal cardinality, the (I, J)-*minor* of M is the determinant
of the square submatrix $M_{IJ} := (m_{ij})_{i \in I, j \in J}$. The *principal minors* are those
with $I = J$ [55, p. 2].) A matrix all of whose minors are positive is said to be
a *P-matrix*.

4.1 Examples

4.1.1 Circuits with ideal diodes

A large amount of electrical network modeling is carried out on the basis of ideal lumped elements: resistors, inductors, capacitors, diodes, and so on. There is not necessarily a one-to-one relation between the elements in a model and the parts of the actual circuit; for instance, a resistor may under some circumstances be better modeled by a parallel connection of an ideal resistor and an ideal capacitor than by an ideal resistor alone. The standard ideal elements should rather be looked at as forming a construction kit from which one can quickly build a variety of models.

Among the standard elements the ideal diode has a special position because of the nonsmoothness of its characteristic. In circuit simulation software that has no ability to cope with mode changes, the ideal diode cannot be admitted as a building block and will have to be replaced for instance by a heavily nonlinear resistor; a price will have to be paid in terms of speed of simulation. The alternative is to work with a hybrid system simulator; more specifically, the software will have to be able to deal with complementarity systems.

To write the equations of a network with (say) k ideal diodes in complementarity form, first extract the diodes so that the network appears as a k-port. For each port, we have a choice between denoting voltage by u_i and current by y_i or vice versa (with the appropriate sign conventions). Often it is possible to make these choices in such a way that the dynamics of the k-port can be written as

$$\dot{x} = f(x, u), \quad y = h(x, u).$$

For linear networks, one can actually show that it is *always* possible to write the dynamics in this form. To achieve this, it may be necessary to let u_i denote voltage at some ports and current at some other ports; in that case one sometimes speaks of a "hybrid" representation, where of course the term is used in a different sense than the one used in this book. Replacing the ports by diodes, we obtain a representation in the semi-explicit complementarity form (4.4).

For electrical networks it is often reasonable to assume that there are no jumps in the continuous state variables, so that there is no need to specify event conditions in addition to the flow conditions (4.4). Complementarity systems in general do not always have continuous solutions, so if one wants to prove that electrical networks with ideal diodes do indeed have continuous solutions, one will have to make a connection with certain specific properties of electrical networks. The passivity property is one that immediately comes to mind, and indeed there are certain conclusions that can be drawn from passivity and that are relevant in the study of properies of complementarity systems. To illustrate this, consider the specific case of a linear passive system coupled to a number of ideal diodes. The system is described by equations of the form

$$\dot{x} = Ax + Bu$$

$$y = Cx + Du \tag{4.6}$$

$$0 \leq y \perp u \geq 0.$$

Under the assumption that the system representation is minimal, the passivity property implies (see [155]) that there exists a positive definite matrix Q such that

$$\begin{bmatrix} A^T Q + QA & QB - C^T \\ B^T Q - C & -(D + D^T) \end{bmatrix} \leq 0. \tag{4.7}$$

Under the condition that the mapping $D + D^T$ is positive definite, one can prove that the complementarity system (4.6) has continuous solutions. If D is equal to zero or more generally is skew-symmetric, then the passivity condition (4.7) implies that $C = B^T Q$, so that in this case the matrix $CB = B^T QB$ is positive definite (assuming that B has full column rank). Under this condition the system (4.6) has solutions with continuous state trajectories if the system is consistently initialized, i. e. the initial condition x_0 satisfies $Cx_0 \geq 0$. See [34] for proofs and additional information on the nature of solutions to linear passive complementarity systems. The importance of the matrices D and CB is related to the fact that they appear in the power series expansion of the transfer matrix $C(sI - A)^{-1}B + D$ around infinity:

$$C(sI - A)^{-1}B + D = D + CBs^{-1} + CABs^{-2} + \cdots .$$

We will return to this when we discuss linear complementarity systems.

4.1.2   Mechanical systems with unilateral constraints

Mechanical systems with geometric inequality constraints (i. e. inequality constraints on the position variables, such as in the simple example of Figure 2.7) are given by equations of the following form (see [138]), in which $\frac{\partial H}{\partial p}$ and $\frac{\partial H}{\partial q}$ denote column vectors of partial derivatives, and the time arguments of q, p, y, and u have been omitted for brevity:

$$\dot{q} = \frac{\partial H}{\partial p}(q, p) \qquad\qquad q \in \mathbb{R}^n, \ p \in \mathbb{R}^n \tag{4.8a}$$

$$\dot{p} = -\frac{\partial H}{\partial q}(q, p) + \frac{\partial C^T}{\partial q}(q)u, \quad u \in \mathbb{R}^k \tag{4.8b}$$

$$y = C(q), \qquad\qquad y \in \mathbb{R}^k \tag{4.8c}$$

$$0 \leq y \perp u \geq 0. \tag{4.8d}$$

Here, $C(q) \geq 0$ is the column vector of geometric inequality constraints, and $u \geq 0$ is the vector of Lagrange multipliers producing the constraint force vector $(\partial C/\partial q)^T(q)u$. (The expression $\partial C^T/\partial q$ denotes an $n \times k$ matrix whose i-th column is given by $\partial C_i/\partial q$.) The perpendicularity condition expresses in particular that the i-th component of u_i can only be non-zero if the i-th constraint is active, that is, $y_i = C_i(q) = 0$. The appearance of the reaction force in the above form, with $u_i \geq 0$, can be derived from the principle that the reaction forces do not exert any work along virtual displacements that are compatible with the constraints. This basic principle of handling geometric inequality constraints can be found e. g. in [125, 86], and dates back to Fourier and Farkas.

The Hamiltonian $H(q,p)$ denotes the total energy, generally given as the sum of a kinetic energy $\frac{1}{2}p^T M^{-1}(q)p$ (where $M(q)$ denotes the mass matrix, depending on the configuration vector q) and a potential energy $V(q)$. The semi-explicit complementarity system (4.8) is called a Hamiltonian complementarity system, since the dynamics of every mode is Hamiltonian [138]. In particular, every mode is energy-conserving, since the constraint forces are workless. It should be noted though that the model could be easily extended to mechanical systems with dissipation by replacing the second set of equations of (4.8) by

$$\dot{p} = -\frac{\partial H}{\partial q}(q,p) - \frac{\partial R}{\partial \dot{q}}(\dot{q}) + \frac{\partial C^T}{\partial q}(q)u \qquad (4.9)$$

where $R(\dot{q})$ denotes a Rayleigh dissipation function.

4.1.3 Optimal control with state constraints

The purpose of this subsection is to indicate in which way one may relate optimal control problems with state constraints to complementarity systems. The study of this subject is far from being complete; we will offer some suggestions rather than present a rigorous treatment. Consider the problem of maximizing a functional of the form

$$\int_0^T F(t, x(t), u(t))dt + F_T(x(T)) \qquad (4.10)$$

over a collection of trajectories described by

$$\dot{x}(t) = f(t, x(t), u(t)), \quad x(0) = x_0 \qquad (4.11)$$

together with the constraints

$$g(t, x(t), u(t)) \geq 0. \qquad (4.12)$$

In the above, g may be a vector-valued function, and then the inequalities are taken componentwise. Under suitable conditions (see [65] for much more detailed information), candidates for optimal solutions can be found by solving a system of equations that is obtained as follows. Let λ be a vector variable of the same length as x, and define the *Hamiltonian* $H(t, x, u, \lambda)$ by

$$H(t, x, u, \lambda) \;=\; F(t, x, u) + \lambda^T f(t, x, u). \tag{4.13}$$

Also, let η be a vector of the same length as g, and define the *Lagrangian* $L(t, x, u, \lambda, \eta)$ by

$$L(t, x, u, \lambda, \eta) \;=\; H(t, x, u, \lambda) + \eta^T g(t, x, u). \tag{4.14}$$

The system referred to before is now the following:

$$\dot{x}(t) \;=\; f(t, x(t), u(t)) \tag{4.15a}$$

$$\dot{\lambda}(t) \;=\; -\frac{\partial L}{\partial x}(t, x(t), u(t), \lambda(t), \eta(t)) \tag{4.15b}$$

$$u(t) \;=\; \arg \max_{\{u \mid g(t, x(t), u) \geq 0\}} L(t, x(t), u, \lambda(t), \eta(t)) \tag{4.15c}$$

$$0 \;\leq\; g(t, x(t), u(t)) \;\perp\; \eta(t) \;\geq\; 0 \tag{4.15d}$$

with initial conditions

$$x(0) = x_0 \tag{4.16}$$

and final conditions

$$\lambda(T) \;=\; \frac{\partial F_T}{\partial x}(x(T)). \tag{4.17}$$

Suppose that $u(t)$ can be solved from (4.15c) so that

$$u(t) \;=\; u^*(t, x(t), \lambda(t), \eta(t)) \tag{4.18}$$

where $u^*(t, x, \lambda, \eta)$ is a certain function. Then define $g^*(t, x, \lambda, \eta)$ by

$$g^*(t, x, \lambda, \eta) \;=\; g(t, x, u^*(t, x, \lambda, \eta)) \tag{4.19}$$

and introduce a variable $y(t)$ by

$$y(t) \;=\; g^*(t, x(t), \lambda(t), \eta(t)). \tag{4.20}$$

The system (4.15) can now be rewritten as

$$\begin{aligned}
\dot{x}(t) &= f(t, x(t), u^*(t, x(t), \lambda(t), \eta(t))) \\
\dot{\lambda}(t) &= -\tfrac{\partial L}{\partial x}(t, x(t), u^*(t, x(t), \lambda(t), \eta(t)), \lambda(t), \eta(t)) \\
y(t) &= g^*(t, x(t), \lambda(t), \eta(t)) \\
0 &\leq y(t) \perp \eta(t) \geq 0.
\end{aligned} \tag{4.21}$$

Here we have a (time-inhomogeneous) complementarity system with state variables x and λ and complementary variables y and η. The system has mixed

boundary conditions (4.16–4.17); therefore one will have existence and uniqueness of solutions under conditions that in general will be different from the ones
that hold for initial-value problems.

A case of special interest is the one in which a quadratic criterion is op-
timized for a linear time-invariant system, subject to linear inequality constraints on the state. Consider for instance the following problem: minimize

$$\tfrac{1}{2} \int_0^T (x(t)^T Q x(t) + u(t)^T u(t)) dt \tag{4.22}$$

subject to

$$\begin{aligned}
\dot{x}(t) &= Ax(t) + Bu(t), \quad x(0) = x_0 \tag{4.23} \\
Cx(t) &\geq 0 \tag{4.24}
\end{aligned}$$

where A, B, and C are matrices of appropriate sizes, and Q is a nonnegative
definite matrix. Following the scheme above leads to the system

$$\begin{aligned}
\dot{x} &= Ax + Bu, \quad x(0) = x_0 \tag{4.25a} \\
\dot{\lambda} &= Qx - A^T \lambda - C^T \eta, \quad \lambda(T) = 0 \tag{4.25b} \\
u &= \arg\max[-\tfrac{1}{2} u^T u + \lambda^T B u] \tag{4.25c} \\
0 &\leq Cx \perp \eta \geq 0 \tag{4.25d}
\end{aligned}$$

where we have omitted the time arguments for brevity. Solving for u from
(4.25c) leads to the equations

$$\frac{d}{dt} \begin{bmatrix} x \\ \lambda \end{bmatrix} = \begin{bmatrix} A & BB^T \\ Q & -A^T \end{bmatrix} \begin{bmatrix} x \\ \lambda \end{bmatrix} + \begin{bmatrix} 0 \\ -C^T \end{bmatrix} \eta \tag{4.26a}$$

$$y = [C \quad 0] \begin{bmatrix} x \\ \lambda \end{bmatrix} \tag{4.26b}$$

$$0 \leq y \perp \eta \geq 0. \tag{4.26c}$$

Not surprisingly, this is a linear Hamiltonian complementarity system.

The study of optimal control problems subject to state constraints is
fraught with difficulties; see Hartl *et al.* [65] for a discussion. The setting
of complementarity systems may be of help in overcoming part of these difficulties.

4.1.4 Variable-structure systems

Consider a nonlinear input-output system of the form

$$\dot{x} = f(x, \bar{u}), \quad \bar{y} = h(x, \bar{u}) \tag{4.27}$$

in which the input and output variables are adorned with a bar for reasons that will become clear in a moment. Suppose that the system is in feedback coupling with a relay element given by

$$
\begin{array}{ll}
\bar{u} = 1, & \bar{y} \geq 0 \\
-1 \leq \bar{u} \leq 1, & \bar{y} = 0 \\
\bar{u} = -1, & \bar{y} \leq 0.
\end{array}
\tag{4.28}
$$

As we have seen above, many of the systems considered in the well-known book by Filippov on discontinuous dynamical systems [54] can be rewritten in this form. At first sight, relay systems do not seem to fit in the complementarity framework. However, let us introduce new variables y_1, y_2, u_1, and u_2, together with the following new equations:

$$
\begin{array}{rcl}
u_1 & = & \frac{1}{2}(1 - \bar{u}) \\
u_2 & = & \frac{1}{2}(1 + \bar{u}) \\
\bar{y} & = & y_1 - y_2
\end{array}
\tag{4.29}
$$

Instead of considering (4.27) together with (4.28), we can also consider (4.27) together with the standard complementarity conditions for the vectors $y = \mathrm{col}(y_1, y_2)$ and $u = \mathrm{col}(u_1, u_2)$:

$$
\left.
\begin{array}{ll}
y_1 = 0, & u_1 \geq 0 \\
y_1 \geq 0, & u_1 = 0
\end{array}
\right.,
\quad
\left.
\begin{array}{ll}
y_2 = 0, & u_2 \geq 0 \\
y_2 \geq 0, & u_2 = 0.
\end{array}
\right.
\tag{4.30}
$$

It can be verified easily that the trajectories of (4.27–4.29–4.30) are the same as those of (4.27–4.28). Note in particular that, although (4.30) in principle allows four modes, the conditions (4.29) imply that $u_1 + u_2 = 1$ so that the mode in which both u_1 and u_2 vanish is excluded, and the actual number of modes is three.

So it turns out that we can rewrite a relay system as a complementarity system, at least if we are willing to accept that some algebraic equations appear in the system description. It is possible to eliminate the variables $\bar{y}$ and $\bar{u}$ and obtain equations in the form

$$
\begin{array}{rcl}
\dot{x} & = & f(x, u_2 - u_1) \\
y_1 - y_2 & = & h(x, u_2 - u_1) \\
u_1 + u_2 & = & 1
\end{array}
\tag{4.31}
$$

together with the complementarity conditions (4.30), but (4.31) is not in standard input-state-output form but rather in a DAE type of form

$$
F(\dot{x}, x, y, u) = 0.
\tag{4.32}
$$

If the relay is a part of a model whose equations are built up from submodels then it is likely anyway that the system description will already be in terms of both differential and algebraic equations, and then it may not be much of a problem to have a few algebraic equations added (depending on how the "index" [28] of the system is affected). Alternatively however one may replace the equations (4.29) by

$$
\begin{aligned}
u_1 &= \tfrac{1}{2}(1 - \bar{u}) \\
y_2 &= \tfrac{1}{2}(1 + \bar{u}) \\
\bar{y} &= y_1 - u_2
\end{aligned}
\tag{4.33}
$$

which are the same as (4.29) except that y_2 and u_2 have traded places. The equations (4.31) can now be rewritten as

$$
\begin{aligned}
\dot{x} &= f(x, 1 - 2u_1) \\
y_1 &= h(x, 1 - 2u_1) + u_2 \\
y_2 &= 1 - u_1
\end{aligned}
\tag{4.34}
$$

and this system does appear in standard input-output form. The only concession one has to make here is that (4.34) will have a feedthrough term (i. e. the output y depends directly on the input u) even when this is not the case in the original system (4.27).

4.1.5 A class of piecewise linear systems

Suppose that a linear system is coupled to a control device which switches between several linear low-level controllers depending on the state of the controlled system, as is the case for instance in many gain scheduling controllers; then the closed-loop system may be described as a piecewise linear system. Another way in which piecewise linear systems may arise is as approximations to nonlinear systems. Modeling by means of piecewise linear systems is attractive because it combines the relative tractability of linear dynamics with a flexibility that is often needed for a precise description of dynamics over a range of operating conditions.

There exist definitions of piecewise linear systems at various levels of generality. Here we shall limit ourselves to systems of the following form (time arguments omitted for brevity):

$$
\begin{aligned}
\dot{x} &= Ax + Bu & \text{(4.35a)} \\
y &= Cx + Du & \text{(4.35b)} \\
(y_i, u_i) &\in \operatorname{graph}(f_i) \quad (i = 1, \ldots, k) & \text{(4.35c)}
\end{aligned}
$$

where, for each i, f_i is a piecewise linear function from $\mathbb{R}$ to $\mathbb{R}^2$. As is common usage, we use the term "piecewise linear" to refer to functions that would in fact be more accurately described as being piecewise affine. We shall consider

functions f_i that are continuous, although from some points of view it would
be natural to include also discontinuous functions; for instance systems in
which the dynamics is described by means of piecewise constant functions
have attracted attention in hybrid systems theory.

The model (4.35) is natural for instance as a description of electrical networks with a number of piecewise linear resistors. Descriptions of this form
are quite common in circuit theory (cf. [93]). Linear relay systems are also
covered by (4.35); note that the "sliding mode" corresponding to the vertical part of the relay characteristic is automatically included. Piecewise linear
friction models are often used in mechanics (for instance Coulomb friction),
which again leads to models like (4.35); see Subsections 2.2.10 and 2.2.11.

One needs to define a solution concept for (4.35); in particular, one has to
say in what function space one will be looking for solutions. With an eye on
the intended applications, it seems reasonable to require that the trajectories
of the variable x should be continuous and piecewise diffentiable. As for the
variable u, some applications suggest that it may also be too much to require
continuity for this variable. For example, take a mass point that is connected
by a linear spring to a fixed wall, and that can move in one direction subject to
Coulomb friction. In a model for this situation the variable u would play the
role of the friction force which, according to the Coulomb model, has constant
magnitude as long as the mass point is moving, and has sign opposite to the
direction of motion. If the mass point is given a sufficiently large initial velocity
away from the fixed wall, it will come to a standstill after some time and then
immediately be pulled back towards the wall, so that in this case the friction
force jumps instantaneously from one end of its interval of possible values to
the other. Even allowing jumps in the variable u, we can still define a solution
of (4.35) to be a triple (x, u, y) such that (4.35b) and (4.35c) hold for almost
all t, and (4.35a) is satisfied in the sense of Carathéodory, that is to say

$$x(t) \;=\; x(0) + \int_0^t [Ax(\tau) + Bu(\tau)]d\tau \tag{4.36}$$

for all t.

The first question that should be answered in connection with the system
(4.35) is whether solutions exist and are unique. For this, one should first of
all find conditions under which, for a given initial condition $x(0) = x_0$, there
exists a unique continuation in one of the possible "modes" of the systems
(corresponding to all possible combinations of the different branches of the
piecewise linear characteristics of the system). This can be a highly nontrivial
problem; for instance in a mechanical system with many friction points, it may
not be so easy to say at which points sliding will take place and at which points
stick will occur. It turns out to be possible to address the problem on the basis
of the theory of the linear complementarity problem and extensions of it. For
the case of Coulomb friction, also in combination with nonlinear dynamics,
this is worked out in [129]. The general case can be developed on the basis
of a theorem by Kaneko and Pang [85], which states that any piecewise linear

characteristic can be described by means of the so-called *Extended Horizontal Linear Complementarity Problem*. On this basis, the piecewise linear system (4.35) may also be described as an extended horizontal linear complementarity system. Results on the solvability of the EHLCP have been given by Sznajder and Gowda [148]. Using these results, one can obtain sufficient conditions for the existence of unique solution starting at a given initial state; see [33] for details.

4.1.6 Projected dynamical systems

The concept of equilibrium is central to mathematical economics. For instance, one may consider an oligopolistic market in which several competitors determine their production levels so as to maximize their profits; it is of interest to study the equilibria that may exist in such a situation. On a wider scale, one may discuss general economic equilibrium involving production, consumption, and prices of commodities. In fact in all kinds of competitive systems the notion of equilibrium is important.

The term "equilibrium" can actually be understood in several ways. For instance, the celebrated Nash equilibrium concept of game theory is defined as a situation in which no player can gain by unilaterally changing his position. Similar notions in mathematical economics lead to concepts of equilibria that can be characterized in terms of systems of *algebraic* equations and inequalities. On the other hand, we have the classical notion of equilibrium in the theory of dynamical systems, where the concept is defined in terms of a given set of *differential* equations. It is natural to expect, though, that certain relations can be found between the static and dynamic equilibrium concepts.

In [49], Dupuis and Nagurney have proposed a general strategy for embedding a given static equilibrium problem into a dynamic system. Dupuis and Nagurney assume that the static equilibrium problem can be formulated in terms of a *variational inequality*; that is to say, the problem is specified by giving a closed convex subset K of $\mathbb{R}^k$ and a function F from K to $\mathbb{R}^k$, and $\bar{x} \in K$ is an equilibrium if

$$\langle F(\bar{x}), x - \bar{x} \rangle \geq 0 \tag{4.37}$$

for all $x \in K$. The formulation in such terms is standard within mathematical programming. With the variational problem they associate a discontinuous dynamical system that is defined by $\dot{x} = -F(x)$ on the interior of K but that is defined differently on the boundary of K in such a way as to make sure that solutions will not leave the convex set K. They then prove that the stationary points of the so defined dynamical system coincide with the solutions of the variational inequality.

In some more detail, the construction proposed by Dupuis and Nagurney can be described as follows. The space $\mathbb{R}^k$ in which state vectors take their values is taken as a Euclidean space with the usual inner product. Let P denote the mapping that assigns to a given point x in $\mathbb{R}^k$ the (uniquely defined) point in K that is closest to x; that is to say,

$$P(x) = \arg \min_{z \in K} ||x - z||. \tag{4.38}$$

For $x \in K$ and a velocity vector $v \in \mathbb{R}^k$, let

$$\pi(x, v) = \lim_{\delta \to 0} \frac{P(x + \delta v) - x}{\delta}. \tag{4.39}$$

If x is in the interior of K, then clearly $\pi(x, v) = v$; however if x is on the boundary of K and v points outwards then $\pi(x, v)$ is a modification of v. The dynamical system considered by Dupuis and Nagurney is now defined by

$$\dot{x} = \pi(x, -F(x)) \tag{4.40}$$

with initial condition x_0 in K. The right hand side of this equation is in general subject to a discontinuous change when the state vector reaches the boundary of K. The state may then follow the boundary along a $(k-1)$-dimensional surface or a part of the boundary characterized by more than one constraint, and after some time it may re-enter the interior of K after which it may again reach the boundary, and so on.

In addition to the expression (4.39) Dupuis and Nagurney also employ a different formulation which has been used in [48]. For this, first introduce the set of inward normals which is defined, for a boundary point x of K, by

$$n(x) = \{\gamma \mid ||\gamma|| = 1, \text{ and } \langle \gamma, x - y \rangle \leq 0, \ \forall y \in K\}. \tag{4.41}$$

If K is a convex polyhedron then the vector defined in (4.39) may equivalently be described by

$$\pi(x, v) = v + \langle v, -\gamma^* \rangle \gamma^* \tag{4.42}$$

where γ^* is defined by

$$\gamma^* := \arg \max_{\gamma \in n(x)} \langle v, -\gamma \rangle. \tag{4.43}$$

A further reformulation is possible by introducing the "cone of admissible velocities". To formulate this concept, first recall that a *curve* in $\mathbb{R}^k$ is a smooth mapping from an interval, say $(-1, 1)$, to $\mathbb{R}^k$. An *admissible velocity* at a point x with respect to the closed convex set $K \subset \mathbb{R}^k$ is any vector that appears as a directional derivative at 0 of a C^∞ function $f(t)$ that satisfies $f(0) = x$ and $f(t) \in K$ for $t \geq 0$. One can show that the set of admissible velocities is a closed convex cone for any x in the boundary of K; of course, the set of admissible velocities is empty when $x \notin K$ and coincides with $\mathbb{R}^k$ if x belongs to the interior of K. One can furthermore show (see [71]) that the mapping defined in (4.42) for given x is in fact just the projection to the cone of admissible velocities. In this way we get an alternative definition of projected dynamical systems. The new formulation is more "intrinsic" in a differential-geometric sense than the original one which is based on the standard coordinatization of k-dimensional Euclidean space. Indeed it would be possible in this way to formulate projected dynamics for systems defined on Riemannian manifolds; the inner product on tangent spaces that is provided by the Riemannian structure makes it possible to define the required projection. One possible application would be the use of projected gradient flows to find minima subject to constraints; cf. for instance [75] for the unconstrained case.

Assume now that the set K is given as an intersection of convex sets of the form $\{x \mid h_i(x) \geq 0\}$ where the functions h_i are smooth. This is actually the situation that one typically finds in applications. It is then possible to reformulate the projected dynamical system as a complementarity system. The construction is described in [71] and we summarize it briefly here. Let $H(x)$ denote the gradient matrix defined by the functions $h_i(x)$; that is to say, the (i, j)-th element of $H(x)$ is

$$(H(x))_{ij} = \frac{\partial h_i}{\partial x_j}(x). \tag{4.44}$$

For $x \in K$, let $I(x)$ be the set of "active" indices, that is,

$$I(x) = \{i \mid h_i(x) = 0\}. \tag{4.45}$$

We denote by $H_{I(x)\bullet}$ the matrix formed by the rows of $H(x)$ whose indices are active; it will be assumed that this matrix has full row rank for all x in the boundary of K ("independent constraints"). One can then show that for each $x \in K$ the cone of admissible velocities is given by $\{v \mid H_{I(x)\bullet}v \geq 0\}$. Moreover, the set of inward normals as defined in (4.41) is given by $\{\gamma \mid ||\gamma|| = 1$ and $\gamma = H_{I(x)\bullet}^T u$ for some $u \geq 0\}$. Consequently, the projection of an arbitrary vector v_0 to the cone of admissible velocities is obtained by solving the minimization problem

$$\min_{v}\{||v_0 - v|| \mid H_{I(x)\bullet}v \geq 0\}.$$

By standard methods, one finds that the minimizer is given by $H_{I(x)\bullet}^T u$ where u is the (unique) solution of the complementarity problem

$$0 \leq H_{I(x)\bullet}v_0 + H_{I(x)\bullet}H_{I(x)\bullet}^T u \perp u \geq 0. \tag{4.46}$$

Now, compare the projected dynamical system (4.40) to the complementarity system defined by

$$\dot{x} = -F(x) + H^T(x)u \tag{4.47a}$$
$$y = h(x) \tag{4.47b}$$
$$0 \leq y \perp u \geq 0 \tag{4.47c}$$

where $h(x)$ is a vector defined in the obvious manner by $(h(x))_i = h_i(x)$, and where the trajectories of all variables are required to be continuous. Suppose that the system is initialized at $t = 0$ at a point x_0 in K. For indices i such that $h_i(x_0) > 0$, the complementarity conditions imply that we must have $u_i(0) = 0$. For indices that are active at x_0 we have $y_i(0) = 0$; to satisfy the inequality constraints also for positive t we need $\dot{y}_i(0) \geq 0$. Moreover, it follows from the complementarity conditions and the continuity conditions that we must have $u_i(0) = 0$ for indices i such that $\dot{y}_i(0) = 0$, and, vice versa, $\dot{y}_i(0) = 0$ for indices i such that $u_i(0) > 0$. Since

$$
\begin{aligned}
\dot{y}_{I(x_0)}(0) &= H_{I(x_0)\bullet}(-F(x_0) + H^T u(0)) \\
&= H_{I(x_0)\bullet}(-F(x_0) + H^T_{I(x_0)\bullet} u_{I(x_0)}(0))
\end{aligned}
$$

the vector $u_{I(x_0)}(0)$ must be a solution of the complementarity problem (4.46). It follows that $H^T u(0)$ is of the form appearing in (4.42). The reverse conclusion follows as well, and moreover one can show that "local" equality of solutions as just shown implies "global" equality [71].

4.1.7   Diffusion with a free boundary

In this subsection we consider a situation in which a complementarity system arises as an approximation to a partial differential equation with a free boundary. We shall take a specific example which arises in the theory of option pricing. For this we first need to introduce some terminology. A *European put option* is a contract that gives the holder the right, but not the obligation, to sell a certain asset to the counterparty in the contract for a specified price (the "exercise price") at a specified time in the future ("time of maturity"). The underlying asset can for instance be a certain amount of stocks, or a certain amount of foreign currency. For a concrete example, consider an investor who has stocks that are worth 100 now and who would like to turn these stocks into cash in one year's time. Of course it is hard to predict what the value of the stocks will be at that time; to make sure that the proceeds will be at least 90, the investor may buy a put option with exercise price 90 that matures in one year. In this way the investor is sure that she can sell the stocks for at least 90.

Of course one has to pay a price to buy such protection, and it is the purpose of option theory to determine "reasonable" option prices. The modern theory of option pricing started in the early seventies with the seminal work by Black, Scholes, and Merton. This theory is not based on the law of large numbers, but rather on the observation that the risk that goes with conferring an option contract is not as big as it would seem to be at first sight. By following an active trading strategy in the underlying asset, the seller ("writer") of the option will be able to reduce the risk. Under suitable model assumptions the risk can even be completely eliminated; that is to say, the cost of providing protection becomes independent of the evolution of the value of the underlying asset and hence can be predicted in advance. The "no-arbitrage" argument then states that this fixed cost must, by the force of competition, be the market price of the option. The model assumptions under which one can show that the risk of writing an option can be eliminated are too strong to be completely realistic; nevertheless, they provide a good guideline for devising strategies that at least are able to reduce risk substantially.

One of the assumptions made in the original work of Black and Scholes [20] is that the price paths of the underlying asset may be described by a stochastic differential equation of the form

$$dS(t) = \mu S(t)dt + \sigma S(t)dw(t) \tag{4.48}$$

where $w(t)$ denotes a standard Wiener process. (See any textbook on SDEs, for instance [122], for the meaning of the above.) Under a number of additional assumptions (for instance: the underlying asset can be traded continuously and without transaction costs, and there is one fixed interest rate r which holds both for borrowing and for lending), Black and Scholes derived a partial differential equation that describes the price of the option at any time before maturity as a function of two variables, namely time t and the price S of the underlying asset. The Black-Scholes equation for the option price $C(S,t)$ is (with omission of the arguments)

$$\frac{\partial C}{\partial t} + \tfrac{1}{2}\sigma^2 S^2 \frac{\partial^2 C}{\partial S^2} + rS\frac{\partial C}{\partial S} - rC = 0 \tag{4.49}$$

with end condition for time of maturity T and exercise price K

$$C(S,T) = \max(K - S, 0) \tag{4.50}$$

and boundary conditions

$$C(0,t) = e^{-r(T-t)}K, \quad \lim_{S\to\infty} C(S,t) = 0. \tag{4.51}$$

It turns out that the "drift" parameter μ in the equation (4.48) is immaterial, whereas the "volatility" parameter σ is very important since it determines the diffusion coefficient in the PDE (4.49).

So far we have been discussing a *European* put option. An *American* put option is the same except that the option may be exercised at any time until the maturity date, rather than only at the time of maturity. (The terms "European" and "American" just serve as a way of distinction; both types of options are traded in both continents.) The possibility of early exercise brings a discrete element into the discussion, since at any time the option may be in two states: "alive" or "exercised". For American options, the Black-Scholes equation (4.49) is replaced by an inequality

$$\frac{\partial C}{\partial t} + \tfrac{1}{2}\sigma^2 S^2 \frac{\partial^2 C}{\partial S^2} + rS\frac{\partial C}{\partial S} - rC \leq 0 \tag{4.52}$$

in which equality holds if the option is not exercised, that is, if its value exceeds the revenues of exercise. For the put option, this is expressed by the inequality

$$C(S,t) > \max(K - S, 0). \tag{4.53}$$

The Black-Scholes equation (4.49) has state-dependent coefficients. A simple substitution, however, will transform it to a constant-coefficient PDE. To this end, express the price of the underlying asset S in terms of a new (dimensionless) independent variable x by

$$S = Ke^x.$$

To make the option price dimensionless as well, introduce $v := C/K$. We also change the final value problem (4.49–4.50) to an initial value problem by setting set $\tau = T - t$. After some computation, we find that the equation (4.49) is replaced by

$$-\frac{\partial v}{\partial \tau} + \tfrac{1}{2}\sigma^2 \frac{\partial^2 v}{\partial x^2} + (r - \tfrac{1}{2}\sigma^2)\frac{\partial v}{\partial x} - rv = 0. \tag{4.54}$$

with the inital and boundary conditions (for the European put option)

$$v(x,0) = \max(1 - e^x, 0), \quad \lim_{x \to -\infty} v(x,\tau) = e^{-r\tau}, \quad \lim_{x \to \infty} v(x,\tau) = 0. \tag{4.55}$$

For the American put option, we get the set of inequalities

$$\frac{\partial v}{\partial \tau} - \tfrac{1}{2}\sigma^2 \frac{\partial^2 v}{\partial x^2} - (r - \tfrac{1}{2}\sigma^2)\frac{\partial v}{\partial x} + rv \geq 0 \tag{4.56}$$

$$v \geq \max(1 - e^x, 0) \tag{4.57}$$

with the boundary conditions

$$\lim_{x \to \infty} v(x,\tau) = 0 \tag{4.58}$$

and

$$v(x,\tau) = 1 - e^x \quad \text{for } x \leq x_f(\tau) \tag{4.59}$$

where $x_f(\tau)$ is the location at time τ of the free boundary which should be determined as part of the problem on the basis of the so-called "smooth pasting" or "high contact" conditions which require that v and $\partial v/\partial x$ should be continuous as functions of x across the free boundary.

Define the function $g(x)$ by

$$g(x) = \max(1 - e^x, 0). \tag{4.60}$$

The partial differential inequality (4.56) and its associated boundary conditions may then be written in the following form which is implicit with respect to the free boundary:

$$(\frac{\partial v}{\partial \tau} - \tfrac{1}{2}\sigma^2 \frac{\partial^2 v}{\partial x^2} - (r - \tfrac{1}{2}\sigma^2)\frac{\partial v}{\partial x} + rv)(v - g) = 0 \tag{4.61a}$$

$$\frac{\partial v}{\partial \tau} - \tfrac{1}{2}\sigma^2 \frac{\partial^2 v}{\partial x^2} - (r - \tfrac{1}{2}\sigma^2)\frac{\partial v}{\partial x} + rv \geq 0, \quad v - g \geq 0. \tag{4.61b}$$

This already suggests a complementarity formulation. Indeed, the above might be considered as an example of an infinite-dimensional complementarity system, both in the sense that the state space dimension is infinite and in the sense that there are infinitely many complementarity relations to be satisfied. A complementarity system of the type that we usually study in this chapter is obtained by approximating the infinite-dimensional system by a finite-dimensional system. For the current application it seems reasonable to carry out the approximation by what is known in numerical analysis as the "method of lines". In this approach there is a first stage in which the space variable is discretized but the time variable is not. Specifically, take a grid of, say, N points in the space variable (in our case this is the dimensionless price variable x), and associate to each grid point x_i a variable $x_i(t)$ which is intended to be an approximation to $v(x_i, t)$. The action of the linear differential operator

$$v \mapsto \tfrac{1}{2}\sigma^2 \frac{\partial^2 v}{\partial x^2} + (r - \tfrac{1}{2}\sigma^2)\frac{\partial v}{\partial x} - rv \tag{4.62}$$

can be approximated by a linear mapping acting on the space spanned by the variables $x_1, \ldots, x_N$. For instance, on an evenly spaced grid an approximation to the first-order differential operator $\partial/\partial x$ is given by

$$A_1 = \frac{1}{2h}\begin{bmatrix} -2 & 2 & 0 & \cdots & \cdots & 0 \\ -1 & 0 & 1 & 0 & & \vdots \\ 0 & -1 & 0 & 1 & \ddots & \vdots \\ \vdots & \ddots & \ddots & \ddots & \ddots & 0 \\ \vdots & & 0 & -1 & 0 & 1 \\ 0 & \cdots & \cdots & 0 & -2 & 2 \end{bmatrix} \tag{4.63}$$

where h is the mesh size, and the second-order differential operator $\partial^2/\partial x^2$ is approximated by

$$A_2 = \frac{1}{h^2}\begin{bmatrix} -2 & 1 & 0 & \cdots & \cdots & 0 \\ 1 & -2 & 1 & 0 & & \vdots \\ 0 & 1 & -2 & 1 & \ddots & \vdots \\ \vdots & \ddots & \ddots & \ddots & \ddots & 0 \\ \vdots & & 0 & 1 & -2 & 1 \\ 0 & \cdots & \cdots & 0 & 1 & -2 \end{bmatrix} \tag{4.64}$$

The mapping (4.62) is then approximated by the matrix

$$A = \tfrac{1}{2}\sigma^2 A_2 + (r - \tfrac{1}{2}\sigma^2)A_1 - rI. \tag{4.65}$$

The function $g(x)$ is represented by the vector g with entries $g_i = g(x_i)$. Consider now the linear complementary system with $N+1$ state variables and N pairs of complementary variables given by

$$\dot{x} = \begin{bmatrix} A & 0 \\ 0 & 0 \end{bmatrix} x + \begin{bmatrix} I \\ 0 \end{bmatrix} u \tag{4.66a}$$

$$y = \begin{bmatrix} I & -g \end{bmatrix} x \tag{4.66b}$$

$$0 \le y \perp u \ge 0. \tag{4.66c}$$

The system is initialized at

$$x_i(0) = g_i \quad (i = 1, \ldots, N), \quad x_{N+1}(0) = 1. \tag{4.67}$$

The complementarity system defined in this way is a natural candidate for providing an approximate solution to the diffusion equation with a free boundary that corresponds to the Black-Scholes equation for an American put option.

A natural idea for generating approximate solutions of complementarity systems is to use an implicit Euler method. For linear complementarity systems of the form

$$\dot{x} = Ax + Bu$$
$$y = Cx + Du \tag{4.68}$$
$$0 \le y \perp u \ge 0$$

the method may be written as follows:

$$\frac{x_{k+1} - x_k}{\Delta t} = Ax_{k+1} + Bu_{k+1} \tag{4.69a}$$

$$y_{k+1} = Cx_{k+1} + Du_{k+1} \tag{4.69b}$$

$$0 \le y_{k+1} \perp u_{k+1} \ge 0 \tag{4.69c}$$

where x_k is intended to be an approximation to $x(k\Delta t)$, and similarly for y and u. At each step this gives rise to a complementarity problem which under suitable assumptions has a unique solution. The results of applying this method above to the equations for an American put option are shown in Fig. 4.1; the figure shows solutions for various times before expiry as a function of the value of the underlying asset (the variable S).

A more standard numerical method for dealing with the American-style Black-Scholes equation is the finite-difference method in which both the time and space variables are discretized; see for instance [158]. In general, an advantage of a "semidiscretization" approach over an approach in which all variables are discretized simultaneously is that one may make use of the highly developed theory of step size control for numerical solutions of ODEs, rather than

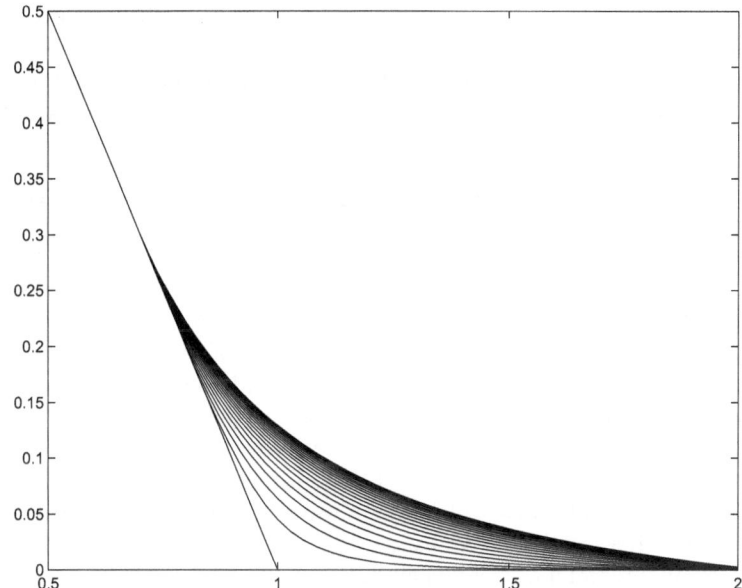

Figure 4.1: Value of an American put option at different times before expiry. Horizontal: current asset value S; vertical: option value C

using a uniform grid. We are of course working here with a complementarity system rather than an ODE, and it must be said that the theory of step size control for complementarity systems is at an early stage of development. Moreover, the theory of approximation of free-boundary problems by complementarity systems has been presented here only for a special case and on the basis of plausibility rather than formal proof, and much further work on this topic is needed.

4.1.8 Max-plus systems

From the fact that the relation

$$z = \max(x, y) \tag{4.70}$$

may also be written in the form

$$z = x + a = y + b, \quad 0 \le a \perp b \ge 0 \tag{4.71}$$

it follows that any system that can be written in terms of linear operations and the "max" operation can also be written as a complementarity system. In particular it follows that the so-called max-plus systems (see [13]), which are closely related to timed Petri nets, can be written as complementarity systems. The resulting equations appear in discrete time, as opposed to the other examples in this section which are all in continuous time; note however that the "time" parameter in a max-plus system is in the standard applications a cycle counter rather than actual time. For further discussion of the relation between the max algebra and the complementarity problem see [44].

4.2 Existence and uniqueness of solutions

Hybrid systems provide a rather wide modeling context, so that there are no easily verifiable necessary and sufficient conditions for well-posedness of general hybrid dynamical systems. It is already of interest to give sufficient conditions for well-posedness of particular classes of hybrid systems, such as complementarity systems. The advantage of considering special classes is that one can hope for conditions that are relatively easy to verify. In a number of special cases, such as mechanical systems or electrical network models, there are moreover natural candidates for such sufficient conditions.

Uniqueness of solutions will below always be understood in the sense of what is sometimes called *right uniqueness*, that is, uniqueness of solutions defined on an interval $[t_0, t_1)$ given an initial state at t_0. It can easily happen in general hybrid systems, and even in complementarity systems, that uniqueness holds in one direction of time but not in the other; take for instance the two-carts example of Subsection 2.2.9 with zero restitution coefficient. We have here one of the points in which discontinuous dynamical systems differ from smooth systems. To allow for the possibility of an initial jump, one may let the initial condition be given at t_0^-.

We have to distinguish between *local* and *global* existence and uniqueness. Local existence and uniqueness, for solutions starting at t_0, holds if there exists an $\varepsilon > 0$ such that on $[t_0, t_0 + \varepsilon)$ there is a unique solution starting at the given initial condition. For global existence and uniqueness, we require that for given initial condition there is a unique solution on $[t_0, \infty)$. If local uniqueness holds for all initial conditions and existence holds globally, then uniqueness must also hold globally since there is no point at which solutions can split. However local existence does not imply global existence. This phenomenon is already well-known in the theory of smooth dynamical systems; for instance the differential equation $\dot{x}(t) = x^2(t)$ with $x(0) = x_0$ has the unique solution $x(t) = x_0(1 - x_0 t)^{-1}$ which for positive x_0 is defined only on the interval $[0, x_0^{-1})$. Some growth conditions have to be imposed to prevent this "escape to infinity". In hybrid systems, there are additional reasons why global existence may fail; in particular we may have an accumulation of mode switches. An example where such an accumulation occurs was already discussed in Chapter 1 (see (1.12)). Another example comes from the so-called *Fuller phenomenon* in optimal control theory [106]. For the problem given by the equations $\dot{x}_1(t) = x_2(t)$, $\dot{x}_2(t) = u(t)$ with end constraint $x(T) = 0$, control constraints $|u(t)| \leq 1$, and cost functional $\int_0^T x_1^2(t)dt$, it turns out that the optimal control is bang-bang (i..e. it takes on only the two extreme values 1 and -1) and has an infinite number of switches. As shown in [90], the phenomenon is not at all rare.

4.3 The mode selection problem

As noted in [138] it is not difficult to find examples of complementarity systems that exhibit nonuniqueness of smooth continuations. For a simple example of this phenomenon within a switching control framework, consider the plant

$$
\begin{aligned}
\dot{x}_1 &= x_2, \qquad y = x_2 \\
\dot{x}_2 &= -x_1 - u
\end{aligned}
\tag{4.72}
$$

in closed-loop with a switching control scheme of relay type

$$
\begin{aligned}
u(t) &= 1, && \text{if} \quad y(t) > 0 \\
-1 &\le u(t) \le 1, && \text{if} \quad y(t) = 0 \\
u(t) &= -1, && \text{if} \quad y(t) < 0.
\end{aligned}
\tag{4.73}
$$

(This could be interpreted as a mass-spring system subject to a "reversed" — and therefore non-physical — Coulomb friction.) It was shown in Subsection 4.1.4 that such a variable-structure system can be modeled as a complementarity system. Note that from any initial (continuous) state $x(0) = (x_1(0), x_2(0)) = (c, 0)$, with $|c| \le 1$, there are three possible smooth continuations for $t \ge 0$ that are allowed by the equations and inequalities above:

(i) $x_1(t) = x_1(0), \quad x_2(t) = 0, \quad u(t) = -x_1(0), \quad -1 \le u(t) \le 1,$
$\quad y(t) = x_2(t) = 0$

(ii) $x_1(t) = -1 + (x_1(0) + 1) \cos t, \quad x_2(t) = -(x_1(0) + 1) \sin t,$
$\quad u(t) = 1, \quad y(t) = x_2(t) < 0$

(iii) $x_1(t) = 1 + (x_1(0) - 1) \cos t, \quad x_2(t) = -(x_1(0) - 1) \sin t,$
$\quad u(t) = -1, \quad y(t) = x_2(t) > 0.$

So the above closed-loop system is not well-posed as a dynamical system. If the sign of the feedback coupling is reversed, however, there is only one smooth continuation from each initial state. This shows that well-posedness is a non-trivial issue to decide upon in a hybrid system, and in particular is a meaningful performance characteristic for hybrid systems arising from switching control schemes.

In this section we follow the treatment of [140] and consider systems of the form

$$
\begin{aligned}
\dot{x}(t) &= f(x(t), u(t)), && x \in \mathbb{R}^n, \quad u \in \mathbb{R}^k \\
y(t) &= h(x(t), u(t)), && y \in \mathbb{R}^k
\end{aligned}
\tag{4.74a}
\tag{4.74b}
$$

with the additional complementarity conditions

$$0 \leq y(t) \perp u(t) \geq 0. \tag{4.75}$$

The functions f and h will always be assumed to be smooth.

The complementarity conditions (4.75) imply that for some index set $I \subset \{1, \ldots, k\}$ one has the algebraic constraints

$$y_i(t) = 0 \ (i \in I), \quad u_i(t) = 0 \ (i \notin I). \tag{4.76}$$

Note that (4.76) always represents k constraints which are to be taken in conjunction with the system of n differential equations in $n + k$ variables appearing in (4.74). The problem of determining which index set I has the property that the solution of (4.74–4.76) coincides with that of (4.74–4.75), at least on an initial time interval, is called the *mode selection problem*. The index set I represents the *mode* of the system.

One approach to solving the mode selection problem would simply be to try all possibilities: solve (4.74) together with (4.76) for some chosen candidate index set I, and see whether the computed solution is such that the inequality constraints $y(t) \geq 0$ and $u(t) \geq 0$ are satisfied on some interval $[0, \varepsilon]$. Under the assumption that smooth continuation is possible from x_0, there must at least be one index set for which the constraints will indeed be satisfied. This method requires in the worst case the integration of 2^k systems of $n + k$ differential/algebraic equations in $n + k$ unknowns.

In order to develop an alternative approach which leads to an *algebraic* problem formulation, let us note first that we can derive from (4.74) a number of relations between the successive time derivatives of $y(\cdot)$, evaluated at $t = 0$, and the same quantities derived from $u(\cdot)$. By successively differentiating (4.74b) and using (4.74a), we get

$$y(t) \quad = \quad h(x(t), u(t)),$$

$$\dot{y}(t) \quad = \quad \frac{\partial h}{\partial x}(x(t), u(t)) f(x(t), u(t)) + \frac{\partial h}{\partial u}(x(t), u(t)) \dot{u}(t)$$

$$=: \quad F_1(x(t), u(t), \dot{u}(t)),$$

and in general

$$y^{(j)}(t) = F_j(x(t), u(t), \ldots, u^{(j)}(t)) \tag{4.77}$$

where F_j is a function that can be specified explicitly in terms of f and h. From the complementarity conditions (4.75), it follows moreover that for each index i either

$$(y_i(0), \dot{y}_i(0), \ldots) = 0 \ \text{ and } \ (u_i(0), \dot{u}_i(0), \ldots) \succeq 0 \tag{4.78}$$

or

$$(y_i(0), \dot{y}_i(0), \ldots) \succeq 0 \ \text{ and } \ (u_i(0), \dot{u}_i(0), \ldots) = 0 \tag{4.79}$$

(or both), where we use the symbol $\succeq$ to denote lexicographic nonnegativity. (A sequence $(a_0, a_1, \ldots)$ of real numbers is said to be *lexicographically non-negative* if either all a_i are zero or the first nonzero element is positive.) This suggests the formulation of the following "dynamic complementarity problem."

Problem DCP. Given smooth functions $F_j : \mathbb{R}^{n+(j+1)k} \to \mathbb{R}^k$ $(j = 0, 1, \ldots,)$ that are constructed from smooth functions $f : \mathbb{R}^n \to \mathbb{R}^n$ and $h : \mathbb{R}^n \to \mathbb{R}^k$ via (4.77), find, for given $x_0 \in \mathbb{R}^n$, sequences $(y^0, y^1, \ldots)$ and $(u^0, u^1, \ldots)$ of k-vectors such that for all j we have

$$y^j = F_j(x_0, u^0, \ldots, u^j) \qquad (4.80)$$

and for each index $i \in \{1, \ldots, k\}$ at least one of the following is true:

$$(y_i^0, y_i^1, \ldots) = 0 \quad \text{and} \quad (u_i^0, u_i^1, \ldots) \succeq 0 \qquad (4.81)$$

$$(y_i^0, y_i^1, \ldots) \succeq 0 \quad \text{and} \quad (u_i^0, u_i^1, \ldots) = 0. \qquad (4.82)$$

We shall also consider truncated versions where j only takes on the values from 0 up to some integer ℓ; the corresponding problem will be denoted by DCP(ℓ). It follows from the triangular structure of the equations that if $((y^0, \ldots, y^\ell), (u^0, \ldots, u^\ell))$ is a solution of DCP(ℓ), then, for any $\ell' < \ell$, $((y^0, \ldots, y^{\ell'}), (u^0, \ldots, u^{\ell'}))$ is a solution of DCP(ℓ'). We call this the *nesting property* of solutions. We define the *active index set at stage* ℓ, denoted by I_ℓ, as the set of indices i for which $(u_i^0, \ldots, u_i^\ell) \succ 0$ in *all* solutions of DCP(ℓ), so that necessarily $y_i^j = 0$ for all j in any solution of DCP (if one exists). Likewise we define the *inactive index set at stage* ℓ, denoted by J_ℓ, as the set of indices i for which $(y_i^0, \ldots, y_i^\ell) \succ 0$ in *all* solutions of DCP(ℓ), so that necessarily $u_i^j = 0$ for all j in any solution of DCP. Finally we define K_ℓ as the complementary index set $\{1, \ldots, k\} \setminus (I_\ell \cup J_\ell)$. It follows from the nesting property of solutions that the index sets I_ℓ and J_ℓ are nondecreasing as functions of ℓ. Since both sequences are obviously bounded above, there must exist an index ℓ^* such that $I_\ell = I_{\ell^*}$ and $J_\ell = J_{\ell^*}$ for all $\ell \geq \ell^*$. We finally note that all index sets defined here of course depend on x_0; we suppress this dependence however to alleviate the notation.

The problem DCP is a generalization of the *nonlinear complementarity problem* (NLCP) (see for instance [39]), which can be formulated as follows: given a smooth function $F : \mathbb{R}^k \to \mathbb{R}^k$, find k-vectors y and u such that $y = F(x, u)$ and $0 \leq y \perp u \geq 0$. For this reason the term "dynamic complementarity problem" as used above seems natural. Apologies are due however to Chen and Mandelbaum who have used the same term in [37] to denote a different although related problem.

Computational methods for the NLCP form a highly active research subject (see [64] for a survey), due to the many applications in particular in equilibrium programming. The DCP is a generalized and parametrized form of the NLCP and given the fact that the latter problem is already considered a major computational challenge, one may wonder whether the approach taken in the previous paragraphs can be viewed as promising from a computational point of view. Fortunately, it turns out that under fairly mild assumptions the DCP can be reduced to a series of *linear* complementarity problems. In the context of mechanical systems this idea is due to Lötstedt [96].

To get a reduction to a sequence of LCPs, assume that the dynamics (4.74) can be written in the *affine* form

$$
\begin{aligned}
\dot{x}(t) &= f(x(t)) + \sum_{i=1}^{k} g_i(x(t)) u_i(t) \\
y(t) &= h(x(t)).
\end{aligned}
\tag{4.83}
$$

Extensive information on systems of this type is given for instance in [121]. In particular we need the following terminology. The *relative degree* of the i-th output y_i is the number of times one has to differentiate y_i to get a result that depends explicitly on the inputs u. The system is said to have *constant uniform relative degree at x_0* if the relative degrees of all outputs are the same and are constant in a neighborhood of x_0.

We can now state the following theorem, in which we use the notation DCP(ℓ) to indicate explicitly the dependence of the dynamic complementarity problem on the number ℓ of differentiation steps. For a proof see [140]. Recall that $L_f h$ denotes the *Lie derivative* of h along the vector field given by f; that is, $L_f h(x) = (\partial h/\partial x)(x) f(x)$. Also, the k-th *Lie derivative* $L_f^k h$ is defined inductively for $k = 2, 3, \ldots$ by $L_f^k h = L_f(L_f^{k-1} h)$ with $L_f^1 h := L_f h$.

Theorem 4.3.1. *Consider the system of equations (4.83) together with the complementarity conditions (4.75), and suppose that the system (4.83) has constant uniform relative degree ρ at a point $x_0 \in \mathbb{R}^n$. Suppose that x_0 is such that*

$$
(h(x_0), \ldots, L_f^{\rho-1} h(x_0)) \succeq 0
\tag{4.84}
$$

(with componentwise interpretation of the lexicographic inequality), and such that all principal minors of the decoupling matrix $L_g L_f^{\rho-1} h(x_0)$ at x_0 are positive. For such x_0, the dynamic complementarity problem DCP(ℓ) has for each ℓ a solution $((y^0, \ldots, y^\ell), (u^0, \ldots, u^\ell))$ which can be found by solving a sequence of LCPs. Moreover this solution is unique, except for the values of u_i^j with $i \notin J_\ell$ and $j > \ell - \rho$.

This result is algebraic in nature. We now return to differential equations; again the proof of the statement below is in [140].

Theorem 4.3.2. *Assume that the functions f, g_i, and h appearing in (4.83) are analytic. Under the conditions of Thm. 4.3.1, there exists an $\varepsilon > 0$ such that (4.83-4.75) has a smooth solution with initial condition x_0 on $[0, \varepsilon]$. Moreover, this solution is unique and corresponds to any mode I such that $I_{\ell^*} \subset I \subset I_{\ell^*} \cup K_{\ell^*}$.*

Example 4.3.3. Consider the following nonlinear complementarity system:

$$\dot{x}_1 = 1 - x_2 u \tag{4.85a}$$
$$\dot{x}_2 = 1 - x_1 u \tag{4.85b}$$
$$y = -x_1 x_2 \tag{4.85c}$$
$$0 \le y \perp u \ge 0. \tag{4.85d}$$

The feasible set consists of the second and fourth quadrants of the (x_1, x_2)-plane. In the mode in which $u = 0$, the dynamics is obviously given by

$$\dot{x}_1 = 1, \quad \dot{x}_2 = 1. \tag{4.86}$$

In the other mode, it follows from $\dot{y} = 0$ that u must satisfy the equation

$$(x_1^2 + x_2^2)u = x_1 + x_2 \tag{4.87}$$

which determines u as a function of x_1 and x_2 except in the origin. The dynamics in the mode $y = 0$ is given for $(x_1, x_2) \neq (0, 0)$ by

$$\dot{x}_1 = 1 - x_2 \frac{x_1 + x_2}{x_1^2 + x_2^2}, \quad \dot{x}_2 = 1 - x_1 \frac{x_1 + x_2}{x_1^2 + x_2^2}. \tag{4.88}$$

However, this can be simplified considerably because when $y = 0$ we must have $x_1 = 0$ or $x_2 = 0$; in the first case we have

$$\dot{x}_1 = 0, \quad \dot{x}_2 = 1 \tag{4.89}$$

and in the second case

$$\dot{x}_1 = 1, \quad \dot{x}_2 = 0. \tag{4.90}$$

The system (4.85a–4.85c) has uniform relative degree 1 everywhere except at the origin, where the relative degree is 3. The decoupling matrix is in this case just a scalar and is given by $x_1^2 + x_2^2$. By Thm. 4.3.2, we find that everywhere except possibly at the origin a unique smooth continuation is possible. The situation at the origin needs to be considered separately. Thm. 4.3.2 does not apply, but we can use a direct argument. Whatever choice we make for u, if $x_1(0) = 0$ and $x_2(0) = 0$ then the equations (4.85a–4.85b) show that $\dot{x}_1(0) = 1$ and $\dot{x}_2(0) = 1$, so that any solution starting from the origin must leave the feasible set. The same conclusion could be drawn from the DCP, since computation shows that for trajectories starting from the origin we have $y(0) = 0$, $\dot{y}(0) = 0$, and $\ddot{y}(0) = -2$ so that the DCP is not solvable. We conclude that the system (4.85) as a whole is not well-posed.

Example 4.3.4 (Passive systems). A system (4.83) is called *passive* (see [155]) if there exists a function $V(x) \ge 0$ (a *storage function*) such that

$$\begin{aligned} &L_f V(x) \le 0 \\ &L_{g_i} V(x) = h_i(x), \quad i = 1, \cdots, k. \end{aligned} \tag{4.91}$$

Let us assume the following non-degeneracy condition on the storage function V:

$$\text{rank } \left[L_{g_j} L_{g_i} V(x)\right]_{i,j=1,\cdots,k} = k, \text{ for all } x \text{ with } h(x) \geq 0. \qquad (4.92)$$

Since $L_{g_j} h_i = L_{g_j} L_{g_i} V$ it follows that the system has uniform relative degree 1, with decoupling matrix $D(x)$ given by the matrix in (4.92). If the principal minors of $D(x)$ are all positive, then well-posedness follows. Note that the condition of $D(x)$ having positive principal minors corresponds to an additional positivity condition on the storage function V. In fact, it can be checked that for a *linear* system with *quadratic* storage function $V(x)$ and no direct feedthrough from inputs to outputs, the decoupling matrix $D(x)$ will be positive definite if $V(x) > 0$ for $x \neq 0$. Hence, if the equations (4.83–4.75) represent a linear passive electrical network containing ideal diodes, then this system is well-posed.

4.4   Linear complementarity systems

4.4.1   Specification

Consider the following system of linear differential and algebraic equations and inequalities

$$\dot{x}(t) \;=\; Ax(t) + Bu(t) \qquad\qquad\qquad (4.93\text{a})$$

$$y(t) \;=\; Cx(t) + Du(t) \qquad\qquad\qquad (4.93\text{b})$$

$$0 \;\leq\; y(t) \;\perp\; u(t) \;\geq\; 0. \qquad\qquad\qquad (4.93\text{c})$$

The equations (4.93a) and (4.93b) constitute a linear system in state space form; the number of inputs is assumed to be equal to the number of outputs. The relations (4.93c) are the usual *complementarity conditions*. The set of indices for which $y_i(t) = 0$ will be referred to as the *active index set*; defining the active index set in this way rather than by the condition $u_i(t) > 0$ used in the previous section allows us to write the inequality constraints corresponding to a given active index set as nonstrict inequalities rather than strict inequalities. The active index set is in general not constant in time, so that the system switches from one "operating mode" to another. To define the dynamics of (4.93) completely, we will have to specify when these mode switches occur, what their effect will be on the state variables, and how a new mode will be selected. A proposal for answering these questions (cf. [69]) will be explained below. The specification of the complete dynamics of (4.93) defines a class of dynamical systems called *linear complementarity systems*.

Let n denote the length of the vector $x(t)$ in the equations (4.93a–4.93b) and let k denote the number of inputs and outputs. There are then 2^k possible choices for the active index set. The equations of motion when the active index set is I are given by

$$
\begin{aligned}
\dot{x}(t) &= Ax(t) + Bu(t) \\
y(t) &= Cx(t) + Du(t) \\
y_i(t) &= 0, \quad i \in I \\
u_i(t) &= 0, \quad i \in I^c
\end{aligned}
\tag{4.94}
$$

where I^c denotes the index set that is complementary to I, that is, $I^c = \{i \in \{1, \ldots, k\} \mid i \notin I\}$. We shall say that the above equations represent the system in *mode I*. An equivalent and somewhat more explicit form is given by the (generalized) state equations

$$
\begin{aligned}
\dot{x}(t) &= Ax(t) + B_{\bullet I} u_I(t) \\
0 &= C_{I \bullet} x(t) + D_{II} u_I(t)
\end{aligned}
\tag{4.95}
$$

together with the output equations

$$
\begin{aligned}
y_{I^c}(t) &= C_{I^c \bullet} x(t) + D_{I^c I} u_I(t) \\
u_{I^c}(t) &= 0.
\end{aligned}
\tag{4.96}
$$

Here and below, the notation $M_{\bullet I}$, where M is a matrix of size $m \times k$ and I is a subset of $\{1, \ldots, k\}$, denotes the submatrix of M formed by taking the columns of M whose indices are in I. The notation $M_{I \bullet}$ denotes the submatrix obtained by taking the rows with indices in the index set I.

Consider the system (4.95) for a given choice of the active index set I. The system does not need in general to have solutions in a classical sense for all possible initial conditions. The initial values of the variable x for which there does exist a continuously differentiable solution are called *consistent states for mode I*. Under conditions that will be specified below, each consistent initial state gives rise to a unique solution (x, u_I) of (4.95). The system (4.93) follows the path of such a solution (it "stays in mode I") as long as the variables $u_I(t)$ defined implicitly by (4.95) and the variables $y_{I^c}(t)$ defined by (4.96) are all nonnegative. As soon as continuation in mode I would lead to a violation of these inequality constraints, a switch to a different mode has to occur. In case the value of the variable $x(t)$ at which violation of the constraints has become imminent is not a consistent state for the new mode, a state jump is called for. So both concerning the dynamics in a given mode and concerning the transitions between different modes there are a number of questions to be answered. For this we shall rely on the geometric theory of linear systems (see [160, 15, 89] for the general background).

Denote by V_I the *consistent subspace* of mode I, i.e. the set of initial conditions x_0 for which there exist smooth functions $x(\cdot)$ and $u_I(\cdot)$, with $x(0) = x_0$, such that (4.95) is satisfied. The space V_I can be computed as the limit of the sequence defined by

$$V_I^0 = \mathbb{R}^n$$

$$V_I^{i+1} = \{x \in V_I^i \mid \exists u \in \mathbb{R}^{|I|} \text{ s.t. } Ax + B_{\bullet I}u \in V_I^i, \ C_{I\bullet}x + D_{II}u = 0\}. \tag{4.97}$$

There exists a linear mapping F_I such that (4.95) will be satisfied for $x_0 \in V_I$ by taking $u_I(t) = F_I x(t)$. The mapping F_I is uniquely determined, and more generally the function $u_I(\cdot)$ that satisfies (4.95) for given $x_0 \in V_I$ is uniquely determined, if the full-column-rank condition

$$\ker \begin{bmatrix} B_{\bullet I} \\ D_{II} \end{bmatrix} = \{0\} \tag{4.98}$$

holds and moreover we have

$$V_I \cap T_I = \{0\}, \tag{4.99}$$

where T_I is the subspace that can be computed as the limit of the following sequence:

$$T_I^0 = \{0\}$$

$$T_I^{i+1} = \{x \in \mathbb{R}^n \mid \exists \tilde{x} \in T_I^i, \ \tilde{u} \in \mathbb{R}^{|I|} \text{ s.t.}$$
$$x = A\tilde{x} + B_{\bullet I}\tilde{u}, \ C_{I\bullet}\tilde{x} + D_{II}\tilde{u} = 0\}. \tag{4.100}$$

As will be indicated below, the subspace T_I is best thought of as the *jump space* associated to mode I, that is, as the space along which fast motions will occur that take an inconsistent initial state instantaneously to a point in the consistent space V_I; note that under the condition (4.99) this projection is uniquely determined. The projection can be used to define a *jump rule*. However, there are 2^k possible projections, corresponding to all possible subsets of $\{1, \ldots, k\}$; which one of these to choose should be determined by a *mode selection rule*.

For the formulation of a mode selection rule we have to relate in some way index sets to continuous states. Such a relation can be established on the basis of the so-called *rational complementarity problem* (RCP). The RCP is defined as follows. Let a rational vector $q(s)$ of length k and a rational matrix $M(s)$ of size $k \times k$ be given. The rational complementarity problem is to find a pair of rational vectors $y(s)$ and $u(s)$ (both of length k) such that

$$y(s) = q(s) + M(s)u(s) \tag{4.101}$$

and moreover for all indices $1 \leq i \leq k$ we have either $y_i(s) = 0$ and $u_i(s) \geq 0$ for all sufficiently large s, or $u_i(s) = 0$ and $y_i(s) \geq 0$ for all sufficiently large s.[1] The vector $q(s)$ and the matrix $M(s)$ are called the *data* of the RCP, and

---

[1]Note the abuse of notation: we use $q(s)$ both to denote the function $s \mapsto q(s)$ and the value of that function at a specific point $s \in \mathbb{C}$. On a few occasions we shall also denote rational functions and rational vectors by single symbols without an argument, which is in principle the proper notation.

we write $\mathrm{RCP}(q(s), M(s))$. We shall also consider an RCP whose data are a quadruple of constant matrices (A, B, C, D) (such as could be used to define (4.93a–4.93b)) and a constant vector x_0, namely by setting

$$q(s) = C(sI - A)^{-1}x_0 \quad \text{and} \quad M(s) = C(sI - A)^{-1}B + D.$$

We say that an index set $I \subset \{1, \ldots, k\}$ *solves* the RCP (4.101) if there exists a solution $(y(s), u(s))$ with $y_i(s) = 0$ for $i \in I$ and $u_i(s) = 0$ for $i \notin I$. The collection of index sets I that solve $\mathrm{RCP}(A, B, C, D; x_0)$ will be denoted by $\mathcal{S}(A, B, C, D; x_0)$ or simply by $\mathcal{S}(x_0)$ if the quadruple (A, B, C, D) is given by the context.

It is convenient to introduce an ordering relation on the field of rational functions $\mathbb{R}(s)$. Given a rational function $f(s)$, we shall say that $f(s)$ is *nonnegative*, and we write $f \succeq 0$, if

$$\exists \sigma_0 \in \mathbb{R} \, \forall \sigma \in \mathbb{R} \, \{\sigma > \sigma_0 \; \Rightarrow \; f(\sigma) \geq 0\}.$$

An ordering relation between rational functions can now be defined by $f \succeq g$ if and only if $f - g \succeq 0$. Note that this defines a *total* ordering on $\mathbb{R}(s)$ so that with this relation $\mathbb{R}(s)$ becomes an ordered field. Extending the conventions that we already have used for the real field we shall say that a rational vector is nonnegative if and only if all its entries are nonnegative, and we shall write $f(s) \perp g(s)$, where $f(s)$ and $g(s)$ are rational vectors, if for each index i at least one of the component functions $f_i(s)$ and $g_i(s)$ is identically zero. With these conventions, the rational complementarity problem may be written in the form

$$
\begin{aligned}
y(s) &= q(s) + M(s)u(s) \\
0 &\preceq y(s) \perp u(s) \succeq 0.
\end{aligned}
\tag{4.102}
$$

After these preparations, we can now proceed to a specification of the complete dynamics of linear complementarity systems. We assume that a quadruple (A, B, C, D) is given whose transfer matrix $G(s) = C(sI - A)^{-1}B + D$ is *totally invertible*, i.e. for each index set I the $k \times k$ matrix $G_{II}(s)$ is nonsingular. Under this condition (see Thm. 4.4.1 below), the two subspaces V_I and T_I as defined above form for all I a direct sum decomposition of the state space $\mathbb{R}^n$, so that the projection along T_I onto V_I is well-defined. We denote this projection by P_I. The interpretation that we give to the equations (4.93) is the following:

$$
\left\|
\begin{aligned}
&\dot{x} = Ax + Bu, \quad y = Cx + Du \\
&u_I \geq 0, \quad y_I = 0, \quad u_{I^c} = 0, \quad y_{I^c} \geq 0 \\
&I^\sharp \in \mathcal{S}(x), \quad x^\sharp = P_{I^\sharp}x.
\end{aligned}
\right.
\tag{4.103}
$$

Below we shall always consider the system (4.93) in the interpretation (4.103).

4.4.2   A distributional interpretation

The interpretation of T_I as a jump space can be made precise by introducing the class of *impulsive-smooth distributions* that was studied by Hautus [66] (see also [67, 57]). The general form of an impulsive-smooth distribution ϕ is

$$\phi = p(\tfrac{d}{dt})\delta + f \tag{4.104}$$

where $p(\cdot)$ is a polynomial, $\frac{d}{dt}$ denotes the distributional derivative, δ is the delta distribution with support at zero, and f is a distribution that can be identified with the restriction to $(0,\infty)$ of some function in $C^\infty(\mathbb{R})$. The class of such distributions will be denoted by C_{imp}. For an element of C_{imp} of the form (4.104), we write $\phi(0^+)$ for the limit value $\lim_{t\downarrow 0} f(t)$. Having introduced the class C_{imp}, we can replace the system of equations (4.95) by its distributional version

$$\begin{aligned}
\tfrac{d}{dt}x &= Ax + B_{\bullet I}u_I + x_0\delta \\
0 &= C_{I\bullet}x + D_{II}u_I
\end{aligned} \tag{4.105}$$

in which the initial condition x_0 appears explicitly, and we can look for a solution of (4.105) in the class of vector-valued impulsive-smooth distributions. It was shown in [67] that if the conditions (4.98) and (4.99) are satisfied, then there exists a unique solution $(x, u_I) \in C_{\text{imp}}^{n+|I|}$ to (4.105) for each $x_0 \in V_I + T_I$; moreover, the solution is such that $x(0^+)$ is equal to $P_I x_0$, the projection of x_0 onto V_I along T_I. The solution is most easily written down in terms of its Laplace transform:

$$\hat{x}(s) = (sI - A)^{-1}x_0 + (sI - A)^{-1}B_{\bullet I}\hat{u}_I(s) \tag{4.106}$$

$$\hat{u}_I(s) = -G_{II}^{-1}(s)C_{I\bullet}(sI - A)^{-1}x_0, \tag{4.107}$$

where

$$G_{II}(s) := C_{I\bullet}(sI - A)^{-1}B_{\bullet I} + D_{II}. \tag{4.108}$$

Note that the notation is consistent in the sense that $G_{II}(s)$ can also be viewed as the (I, I)-submatrix of the transfer matrix $G(s) := C(sI - A)^{-1}B + D$. It is shown in [67] (see also [117]) that the transfer matrix $G_{II}(s)$ associated to the system parameters in (4.95) is left invertible when (4.98) and (4.99) are satisfied. Since the transfer matrices $G_{II}(s)$ that we consider are square, left invertibility is enough to imply invertibility, and so (by duality) we also have $V_I + T_I = \mathbb{R}^n$. Summarizing, we can list the following equivalent conditions.

Theorem 4.4.1. *Consider a time-invariant linear system with k inputs and k outputs, given by standard state space parameters (A, B, C, D). The following conditions are equivalent.*

    1. For each index set $I \subset \bar{k}$, the associated system (4.95) admits for each $x_0 \in V_I$ a unique smooth solution (x, u) such that $x(0) = x_0$.

2. *For each index set $I \subset \bar{k}$, the associated distributional system (4.105) admits for each initial condition x_0 a unique impulsive-smooth solution (x, u).*

3. *The conditions (4.98) and (4.99) are satisfied for all $I \subset \bar{k}$.*

4. *The transfer matrix $G(s) = C(sI - A)^{-1}B + D$ is totally invertible (as a matrix over the field of rational functions).*

In connection with the system (4.93) it makes sense to introduce the following definitions.

Definition 4.4.2. An impulsive-smooth distribution $\phi = p(\frac{d}{dt})\delta + f$ as in (4.104) will be called *initially nonnegative* if the leading coefficient of the polynomial $p(\cdot)$ is positive, or, in case $p = 0$, the smooth function f is non-negative on some interval of the form $(0, \varepsilon)$ with $\varepsilon > 0$. A vector-valued impulsive-smooth distribution will be called *initially nonnegative* if each of its components is initially nonnegative in the above sense.

Definition 4.4.3. A triple of vector-valued impulsive-smooth distributions (u, x, y) will be called an *initial solution* to (4.93) with *initial state* x_0 and *solution mode I* if

1. the triple (u, x, y) satisfies the distributional equations

$$\frac{d}{dt}x = Ax + Bu + x_0\delta$$
$$y = Cx + Du$$

2. both u and y are initially nonnegative

3. $y_i = 0$ for all $i \in I$ and $u_i = 0$ for all $i \notin I$.

For an impulsive-smooth distribution w that has a rational Laplace transform $\hat{w}(s)$ (such as in (4.106) and (4.107)), we have that w is initially nonneg-ative if and only if $\hat{w}(s)$ is nonnegative for all sufficiently large real values of s. From this it follows that the collection of index sets I for which there exists an initial solution to (4.93) with initial state x_0 and solution mode I is exactly $\mathcal{S}(x_0)$ as we defined this set before in terms of the rational complementarity problem.

An alternative approach to the construction of initial solutions for linear complementarity systems proceeds through the linear version of the dynamic complementarity problem that was discussed in Section 4.3. It has been shown by De Schutter and De Moor that the dynamic linear complementarity prob-lem can be rewritten as a version of the LCP known as the "extended linear complementarity problem" (see [45] for details).

4.4.3   Well-posedness

Most of the well-posedness results that are available for linear complementarity systems provide only sufficient conditions. For the case of *bimodal* linear complementarity systems, i.e. systems with only two modes ($k = 1$), well-posedness has been completely characterized however (see [73]; cf. also [138] for an earlier result with a slightly different notion of well-posedness). Note that a system of the form (4.93a-4.93b) has a transfer function $g(s) = C(sI - A)^{-1}B + D$ which is a rational function. In this case the conditions of Thm. 4.4.1 apply if $g(s)$ is nonzero. The *Markov parameters* of the system are the coefficients of the expansion of $g(s)$ around infinity,

$$g(s) = g_0 + g_1 s^{-1} + g_2 s^{-2} + \cdots .$$

The *leading Markov parameter* is the first parameter in the sequence g_0, g_1, ... that is nonzero. In the theorem below it is assumed that the output matrix C is nonzero; note that if $C = 0$ and $D \neq 0$ the system (4.93) is just a complicated way of representing the equations $\dot{x} = Ax$, so that in that case we do not really have a bimodal system.

Theorem 4.4.4. *Consider the linear complementarity system (4.93) under the assumptions that $k = 1$ (only one pair of complementary variables) and $C \neq 0$; also assume that the transfer function $g(s) = C(sI - A)^{-1}B + D$ is not identically zero. Under these conditions, the system (4.93) has for all initial conditions a unique piecewise differentiable right-Zeno solution if and only if the leading Markov parameter is positive.*

It is typical to find that well-posedness of complementarity systems is linked to a positivity condition. If the number of pairs of complementary variables is larger than one, an appropriate matrix version of the positivity condition has to be used. As might be expected, the type of positivity that we need is the "P-matrix" property from the theory of the LCP. Recall (see the end of the Introduction of this chapter) that a square real matrix is said to be a P-matrix if all its principal minors are positive.

To state a result on well-posedness for multivariable linear complementarity systems, we again need some concepts form linear system theory. Recall (see for instance [84, p. 384] or [89, p. 24]) that a square rational matrix $G(s)$ is said to be *row proper* if it can be written in the form

$$G(s) = \Delta(s)B(s) \tag{4.109}$$

where $\Delta(s)$ is a diagonal matrix whose diagonal entries are of the form s^k for some integer k that may be different for different entries, and $B(s)$ is a proper rational matrix that has a proper rational inverse (i.e. $B(s)$ is *bicausal*). A proper rational matrix $B(s) = B_0 + B_1 s^{-1} + \cdots$ has a proper rational inverse if and only if the constant matrix B_0 is invertible. This constant matrix is uniquely determined in a factorization of the above form; it is called the *leading row coefficient matrix* of $G(s)$. In a completely similar way one defines the notions of column properness and of the leading column coefficient matrix. We can now state the following result [69, Thm. 6.3].

Theorem 4.4.5. *The linear complementarity system (4.93) is well-posed if the associated transfer matrix $G(s) = C(sI - A)^{-1}B + D$ is both row and column proper, and if both the leading row coefficient matrix and the leading column coefficient matrix are P-matrices. Moreover, in this case the multiplicity of events is at most one, i. e. at most one reinitialization takes place at times when a mode change occurs.*

An alternative sufficient condition for well-posedness can be based on the rational complementarity problem (RCP) that was already used above (section 4.4). For a given set of linear system parameters (A, B, C, D), we denote by $\text{RCP}(x_0)$ the rational complementarity problem $\text{RCP}(q(s), M(s))$ with data $q(s) = C(sI - A)^{-1}x_0$ and $M(s) = C(sI - A)^{-1}B + D$. For the purposes of simplicity, the following result is stated under somewhat stronger hypotheses than were used in the original paper [70, Thm. 5.10, 5.16].

Theorem 4.4.6. *Consider the linear complementarity system (4.93), and assume that the associated transfer matrix is totally invertible. The system (4.93) is well-posed if the problem $\text{RCP}(x_0)$ has a unique solution for all x_0.*

A connection between the rational complementarity problem and the standard linear complementarity problem can be established in the following way [70, Thm. 4.1, Cor. 4.10].

Theorem 4.4.7. *For given $q(s) \in \mathbb{R}^k(s)$ and $M(s) \in \mathbb{R}^{k \times k}(s)$, the problem $\text{RCP}(q(s), M(s))$ is uniquely solvable if and only if there exists $\mu \in \mathbb{R}$ such that for all $\lambda > \mu$ the problem $\text{LCP}(q(\lambda), M(\lambda))$ is uniquely solvable.*

The above theorem provides a convenient way of proving well-posedness for several classes of linear complementarity systems. The following example is taken from [70].

Example 4.4.8. A linear mechanical system may be described by equations of the form

$$M\ddot{q} + D\dot{q} + Kq = 0 \tag{4.110}$$

where q is the vector of generalized coordinates, M is the generalized mass matrix, D contains damping and gyroscopic terms, and K is the elasticity matrix. The mass matrix M is positive definite. Suppose now that we subject the above system to unilateral constraints of the form

$$Fq \geq 0 \tag{4.111}$$

where F is a given matrix. Under the assumption of inelastic collisions, the dynamics of the resulting system may be described by

$$M\ddot{q} + D\dot{q} + Kq = F^T u, \quad y = Fq \tag{4.112}$$

together with complementarity conditions between y and u. The associated RCP is the following:

$$y(s) = F(s^2 M + sD + K)^{-1}[(sM + D)q_0 + M\dot{q}_0]$$
$$+ F(s^2 M + sD + K)^{-1} F^T u(s). \quad (4.113)$$

If F has full row rank, then the matrix $F(s^2 M + sD + K)^{-1} F^T$ is positive definite (although not necessarily symmetric) for all sufficiently large s because the term associated to s^2 becomes dominant. By combining the standard result on solvability of LCPs with Thm. 4.4.7, it follows that RCP is solvable and we can use this to prove the well-posedness of the constrained mechanical system. This provides some confirmation for the validity of the model that has been used, since physical intuition certainly suggests that a unique solution should exist.

In the above example, one can easily imagine cases in which the matrix F does not have full row rank so that the fulfillment of some constraints already implies that some other constraints will also be satisfied; think for instance of a chair having four legs on the ground. In such cases the basic result on solvability of LCPs does not provide enough information, but there are alternatives available that make use of the special structure that is present in equations like (4.113). On the basis of this, one can still prove well-posedness; in particular the trajectories of the coordinate vector $q(t)$ are uniquely determined, even though the trajectories of the constraint force $u(t)$ are not.

4.5   Mechanical complementarity systems

Mechanical systems with unilateral constraints can be represented as semi-explicit complementarity systems (cf. [138]):

$$\dot{q} = \frac{\partial H}{\partial p}(q,p) \qquad\qquad q \in \mathbb{R}^n, p \in \mathbb{R}^n$$
$$\dot{p} = -\frac{\partial H}{\partial q}(q,p) - \frac{\partial R}{\partial \dot{q}}(\dot{q}) + \frac{\partial C^T}{\partial q}(q)u \quad u \in \mathbb{R}^k \qquad (4.114a)$$
$$y = C(q), \qquad\qquad\qquad y \in \mathbb{R}^k$$

$$0 \leq y \perp u \geq 0. \qquad\qquad\qquad\qquad\qquad (4.114b)$$

The presentation here is the same as in (4.8) except that we have added a Rayleigh dissipation function R. Assume that the system (4.114a) is real-analytic, and that the unilateral constraints are *independent*, that is

$$\text{rank } \frac{\partial C^T}{\partial q}(q) = k, \text{ for all } q \text{ with } C(q) \geq 0. \qquad (4.115)$$

Since the Hamiltonian is of the form (kinetic energy plus potential energy)

$$H(q,p) = \frac{1}{2}p^T M^{-1}(q)p + V(q), \quad M(q) = M^T(q) > 0 \qquad (4.116)$$

where $M(q)$ is the generalized mass matrix, it follows that the system (4.114a) has uniform relative degree 2 with decoupling matrix

$$D(q) = \left[\frac{\partial C^T}{\partial q}(q) \right]^T M^{-1}(q) \frac{\partial C^T}{\partial q}(q). \qquad (4.117)$$

Hence, from $M(q) > 0$ and (4.115) it follows that $D(q)$ is *positive definite* for all q with $C(q) \geq 0$. Since the principal minors of a positive definite matrix are all positive, all conditions of Thm. 4.3.1 and Thm. 4.3.2 are satisfied, and we establish well-posedness for smooth continuations. We have essentially followed an argument in [96].

A switch rule for mechanical complementarity systems can be formulated as follows. Let us consider a mechanical system with n degrees of freedom $q = (q_1, \cdots, q_n)$ having kinetic energy $\frac{1}{2} \dot{q}^T M(q) \dot{q}$, where $M(q) > 0$ is the generalized mass matrix. Suppose the system is subject to k geometric inequality constraints

$$y_i = C_i(q) \geq 0, \quad i \in K = \{1, \cdots, k\} \qquad (4.118)$$

If the i-th inequality constraint is *active*, that is $C_i(q) = 0$, then the system will experience a constraint force of the form $\frac{\partial C_i}{\partial q}(q) u_i$, where $\frac{\partial C_i}{\partial q}(q)$ is the column vector of partial derivatives of C_i and u_i a Lagrange multiplier.

Let us now consider an arbitrary initial continuous state $(q^-, \dot{q}^-)$. Define the vector of generalized velocities

$$v^- := \frac{\partial C_I}{\partial q}(q^-)\dot{q}^- \qquad (4.119)$$

where I denotes the set of active indices at q^-. In order to describe the inelastic collision we consider the system of equalities and inequalities (in the unknowns v^+, λ)

$$v^+ = v^- + \frac{\partial C_I}{\partial q}(q^-)M^{-1}(q^-)\frac{\partial C_I^T}{\partial q}(q^-)\lambda$$
$$0 \leq v^+ \perp \lambda \geq 0. \qquad (4.120)$$

Here λ can be interpreted as a vector of Lagrange multipliers related to *impulsive forces*. The system (4.120) is in the form of a *linear complementarity problem* (LCP). The general form of an LCP can be written as

$$y = x + Mu, \quad 0 \leq y \perp u \geq 0 \qquad (4.121)$$

where the vector x and the square matrix M are given, and the vectors y and u are the unknowns. As already noted in the Introduction of this chapter, it is a classical result that the LCP (4.121) has a unique solution (y, u) for each x if and only if the matrix M is a P-matrix, that is to say, if and only if all principal minors of the matrix M are positive. This holds in particular if M

is a positive definite matrix. Since $\frac{\partial C_I}{\partial q}(q^-)M^{-1}(q^-)\frac{\partial C_I^T}{\partial q}(q^-) > 0$, the LCP (4.120) has a unique solution. The *jump rule* is now given by

$$(q^-, \dot{q}^-) \mapsto (q^+, \dot{q}^+),$$

with

$$q^+ = q^-, \quad \dot{q}^+ = \dot{q}^- + M^{-1}(q^-)\frac{\partial C_I^T}{\partial q}(q^-)\lambda. \tag{4.122}$$

The new velocity vector $\dot{q}^+$ may equivalently be characterized as the solution of the quadratic programming problem

$$\min_{\{\dot{q}^+|C(q)\dot{q}^+\geq 0\}} \tfrac{1}{2}(\dot{q}^+ - \dot{q}^-)^T M(q)(\dot{q}^+ - \dot{q}^-) \tag{4.123}$$

where $q := q^- = q^+$. This formulation is sometimes taken as a basic principle for describing multiple inelastic collisions; see [31, 112]. Note that there is a simple interpretation to the quadratic programming problem (4.123): in the tangent space at the configuration q, the problems calls for the determination of the admissible velocity that is closest to the impact velocity, where "closest" is interpreted in the sense of the metric given by the kinetic energy. An appealing feature of the transition rule above is that the energy of the mechanical system will always decrease at the switching instant. One may take this as a starting point for stability analysis.

Example 4.5.1 (Two carts with stop and hook). As an example of the switching rule described above, let us again consider the two-carts system of Subsection 2.2.9. To make things more interesting, we add a second constraint which might be realized as a hook (see Fig. 4.2). The equations of motion are

$$\ddot{x}_1(t) = -2x_1(t) + x_2(t) + u_1(t) + u_2(t) \tag{4.124a}$$
$$\ddot{x}_2(t) = x_1(t) - x_2(t) - u_2(t) \tag{4.124b}$$
$$y_1(t) = x_1(t) \tag{4.124c}$$
$$y_2(t) = x_1(t) - x_2(t) \tag{4.124d}$$
$$0 \leq y_1(t) \perp u_1(t) \geq 0 \tag{4.124e}$$
$$0 \leq y_2(t) \perp u_2(t) \geq 0. \tag{4.124f}$$

Consider now the multiple-impact point $(x_1, x_2) = (0, 0)$. The two-dimensional tangent space at this point contains several regions of interest which are indicated in Fig. 4.3. In the figure, the horizontal axis is used for $v_1 := \dot{x}_1$ and the vertical axis for $v_2 := \dot{x}_2$. The cone of admissible post-impact velocities is given by the conditions $v_1 \geq 0$ and $v_1 - v_2 \geq 0$; this region has been labeled A in the figure. The opposite cone contains all possible pre-impact velocities, and consists of three regions which have been labeled B, C and D. According to the jump rule specified above, a given pre-impact velocity will be mapped to the post-impact velocity that is closest to it in the sense of the

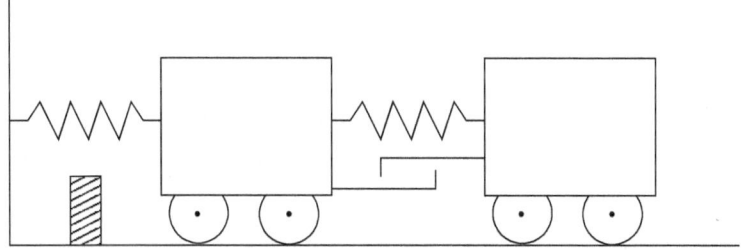

Figure 4.2: Two carts with stop and hook

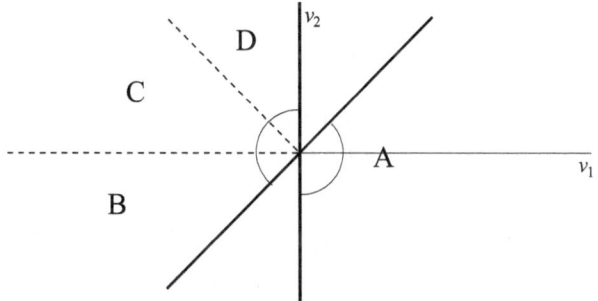

Figure 4.3: Tangent plane

kinetic metric, which in this case is just the standard Euclidean metric since the masses of both carts are assumed to be 1. (We leave it to the reader to work out the jump condition in other cases, for instance when the mass of the right cart is twice the mass of the left cart.) Therefore, a pre-impact velocity in region B will be mapped to a post-impact velocity on the halfline $v_1 = 0$, $v_2 \leq 0$, so that the left cart will remain in contact with the stop whereas the hook contact is not active. For pre-impact velocities in region C, the origin is the closest point in the cone of admissible velocities. This means that if the pre-impact velocities are such that $v_1 \leq 0$ and $0 \leq v_2 < -v_1$, the impact will bring the system to an instantaneous standstill. Finally if $v_1 \leq 0$ and $v_2 \geq -v_1$ just before impact (region D), then the hook contact will be maintained, so the two carts remain at a constant distance, whereas the contact at the stop is released (except in the trivial case in which the pre-impact velocities are zero, so that in fact there is no impact at all).

Since the system in our example is linear, we may also treat it by the methods of Section 4.4; in particular we may apply the RCP-based switching rule (4.103). In the present case, the rational complementarity problem (4.102) takes the following form, for general initial conditions $x_i(0) = x_{i0}$, $\dot{x}_i(0) = v_{i0}$

$(i = 1, 2)$:

$$(s^4 + 3s^2 + 1)y_1(s) =$$
$$= (s^2 + 1)sx_{10} + sx_{20} + (s^2 + 1)v_{10} + v_{20}+$$
$$+ (s^2 + 1)u_1(s) + s^2 u_2(s) \quad (4.125a)$$

$$(s^4 + 3s^2 + 1)y_2(s) =$$
$$= s^3 x_{10} - (s^2 + 1)sx_{20} + s^2 v_{10} - (s^2 + 1)v_{20}+$$
$$+ s^2 u_1(s) + (2s^2 + 1)u_2(s). \quad (4.125b)$$

We are interested in particular in the situation in which $x_{10} = 0$ and $x_{20} = 0$. To find out under what conditions the rule (4.103) allows for instance a jump to the stop-constrained mode, we have to solve the above rational equations for $y_2(s)$ and $u_1(s)$ under the conditions $y_1(s) = 0$ and $u_2(s) = 0$. The resulting equations are

$$0 = (s^2 + 1)v_{10} + v_{20} + (s^2 + 1)u_1(s) \quad (4.126a)$$
$$(s^4 + 3s^2 + 1)y_2(s) = s^2 v_{10} - (s^2 + 1)v_{20} + s^2 u_1(s). \quad (4.126b)$$

These equations can be readily solved, and one obtains

$$y_2(s) = -\frac{1}{s^2 + 1}v_{20} \quad (4.127a)$$
$$u_1(s) = -v_{10} - \frac{1}{s^2 + 1}v_{20}. \quad (4.127b)$$

These functions are nonnegative in the ordering we have put on rational functions if and only if $v_{10} \leq 0$ and $v_{20} \leq 0$; so it follows that the jump to the stop-constrained mode is possible according to the RCP rule if and only if these conditions are satisfied. Moreover, the re-initialization that takes place is determined by (4.127b) and consists of the mapping $(v_{10}, v_{20}) \mapsto (0, v_{20})$. Note that these results are in full agreement with the projection rule based on the kinetic metric. This concludes the computations for the jump to the stop-constrained mode. In the same way one can verify that actually in all cases the RCP rule leads to the same results as the projection rule. It can be proved for linear Hamiltonian complementarity systems in general that the RCP rule and the projection rule lead to the same jumps; see [69].

It should be noted that in the above example we consider only some of the first elements of impact theory. In applied mechanics one needs to deal with much more complicated impacts in which also frictional effects and elastic deformations may play a role. In any given particular situation, one needs to look for impact models that contain enough freedom to allow a satisfactory description of a range of observed phenomena, and that at the same time are reasonably identifiable in the sense that parameter values can be obtained to good accuracy from experiments. In addition to the modeling problems, one also faces very substantial computational problems in situations where there are many contact points, such as for instance in the study of granular material.

Remark 4.5.2. If in the example above one replaces the initial data $(x_{10}, x_{20}) = (0, 0)$ by $(x_{10}, x_{20}) = (\varepsilon, 0)$ or $(x_{10}, x_{20}) = (0, -\varepsilon)$, then one obtains quite different solutions. The system in the example therefore displays *discontinuous dependence on inital conditions*. We see that such a phenomenon, which is quite rare for ordinary differential equations, occurs already in quite simple examples of hybrid dynamical systems. This again indicates that there are fundamental diffences between smooth dynamical systems and hybrid dynamical systems; we already noted before that in simple examples of hybrid systems one may have right uniqueness of solutions but no left uniqueness, which is also a phenomenon that normally doesn't occur in systems described by ordinary differential equations.

The discontinuous dependence on initial conditions may be viewed as a result of an idealization, reflecting a very sensitive dependence on initial conditions in a corresponding smooth model. Certainly the fact that such discontinuities appear is a problem in numerical simulation, but numerical problems would also occur when for instance the strict unilateral constraints in the example above would be replaced by very stiff springs. So the hybrid model in itself cannot be held responsible for the (near-)discontinuity problem; one should rather say that it brings this problem to light.

4.6 Relay systems

For piecewise linear relay systems of the form

$$\dot{x} = Ax + Bu, \quad y = Cx + Du, \quad u_i = -\operatorname{sgn} y_i \ (i = 1, \ldots, k) \qquad (4.128)$$

one may apply Thm. 4.4.7, but the application is not straightforward for the following reason. As noted above, it is possible to rewrite a relay system as a complementarity system (in several ways actually). Using the method (4.29), one arrives at a relation between the new inputs $\operatorname{col}(u_1, u_2)$ and the new outputs $\operatorname{col}(y_1, y_2)$ that may be written in the frequency domain as follows ($\mathbb{1}$ denotes the vector all of whose entries are 1, and $G(s)$ denotes the transfer matrix $C(sI - A)^{-1}B + D$):

$$\begin{bmatrix} u_1(s) \\ u_2(s) \end{bmatrix} = \begin{bmatrix} -G^{-1}(s)C(sI - A)^{-1}x_0 + s^{-1}\mathbb{1} \\ G^{-1}(s)C(sI - A)^{-1}x_0 + s^{-1}\mathbb{1} \end{bmatrix} + $$
$$+ \begin{bmatrix} G^{-1}(s) & -G^{-1}(s) \\ -G^{-1}(s) & G^{-1}(s) \end{bmatrix} \begin{bmatrix} y_1(s) \\ y_2(s) \end{bmatrix}. \qquad (4.129)$$

The matrix that appears on the right hand side is singular for all s and so the corresponding LCP does not always have a unique solution. However the expression at the right hand side is of a special form and we only need to ensure existence of a unique solution for LCPs of this particular form. On the basis of this observation, the following result is obtained.

Theorem 4.6.1. [95, 70] *The piecewise linear relay system (4.128) is well-posed if the transfer matrix $G(s)$ is a P-matrix for all sufficiently large s.*

In particular the theorem may be used to verify that the system in the example at the beginning of Section 4.3, with the "right" sign of the relay feedback, is well-posed.

The above result gives a criterion that is straightforward to verify (compute the determinants of all principal minors of $G(s)$, and check the signs of the leading Markov parameters), but that is restricted to piecewise linear systems. Filippov [54, §2.10] gives a criterion for well-posedness which works for general nonlinear systems, but needs to be verified on a point-by-point basis.

4.7   Notes and references for Chapter 4

Complementarity problems have been studied intensively since the mid-sixties, and much attention has in particular been given to the linear complementarity problem. The book [39] by Cottle, Pang, and Stone provides a rich source of information on the LCP. The general formulation of input-output dynamical systems as in (4.4a–4.4b) has been popular in particular in control theory since the early sixties. See for instance [121] for a discussion of nonlinear systems, and [84] for linear systems. The combination of complementarity conditions and differential equations has been used for mechanical problems by Lötstedt [96]; related work has been done by Moreau who used a somewhat different formulation (the "sweeping process") [113]. In the mechanical context, complementarity conditions have not only been used for the description of unilateral constraints but also for the modeling of dry friction; see for instance [129, 146, 145]. Another area where the combination of complementarity conditions and differential equations arises naturally is electrical circuit simulation; early work in the simulation of piecewise linear electrical networks has been done by Van Bokhoven [23]. The idea of combining complementarity conditions with general input-output dynamical systems seems to have been proposed first in [138].[2] Further information about complementarity systems is available in [33, 34, 68, 69, 70, 71, 72, 73, 95, 139, 140, 141].

---

[2]In the cited paper the term "complementary-slackness system" was used. In later work this has been changed to "complementarity system" because this term is shorter and connects more closely to the complementarity problem, and also because the English word *slackness* seems to be hard to translate to other languages such as Dutch.

Chapter 5

Analysis of hybrid systems

Typical properties studied for smooth dynamical systems include the nature of equilibria, stability, reachability, the presence of limit cycles, and the effects of parameter changes. In the domain of finite automata, one may be interested for instance in the occurrence of deadlock, the correctness of programs, inclusion relations between languages recognized by different automata, and issues of complexity and decidability. In the context of hybrid systems, one may expect to encounter properties both from the continuous and from the discrete domain, with some interaction between them. In this chapter we discuss a number of examples.

5.1 Correctness and reachability

Over the past three centuries, the investigation of properties of smooth dynamical systems has been a subject of intense research. Ingenious methods have been used to obtain many results of interest and of practical relevance, but of course the subject is too wide to be ever completely understood. The typical approach is to look for properties that can be proven to hold for suitably defined subclasses; there is no general proof method, and usually considerable ingenuity is needed to arrive at interesting conclusions. In the theory of finite automata however we encounter a situation that is radically different. Since the number of states is finite, certain properties can be verified in principle by checking all possible states. The catch is in the words "in principle"; the number of states to visit may indeed be huge in situations of practical interest, and so one should be concerned with the complexity of algorithms and various heuristics.

5.1.1 Formal verification

The term *formal verification* (or *computer-aided verification*) is used for the methods that have been developed in computer science to analyze the properties of discrete systems. Of particular interest is to prove the *correctness* of programs (understood in a wide sense to include software but also communication protocols and hardware descriptions). To find bugs in a complex software system is a highly nontrivial matter, as is well-known, and given the size of the software industry there is a major interest in automated tools that can help in the verification process.

There are two types of formal verification methods, known as *theorem proving* and *model checking*. In theorem proving, one starts with a representation of the program to be analyzed and attempts to arrive at a proof of correctness by applying a set of inference rules. Theorem provers often need some human guidance to complete their task in a reasonable amount of time. Moreover, when the proof of correctness fails, a theorem prover often provides little indication of what is wrong in the design. Model checking is in principle a brute-force approach, building on the finiteness of the state space. By searching all possibilities, the method either proves a design to be correct or finds a counterexample. The ability of model checkers to find bugs is of course very important to system designers. The time needed to do an exhaustive search is however exponential in terms of the number of variables in the system, and so considerable attention has gone into the development of methods that can combat complexity.

Program properties to be specified are often expressed in terms of *temporal logic* formulas; these are able to express properties like "if the variable P is reset to zero, then eventually the variable Q is reset to zero as well". More generally, correctness can be viewed as an *implementation relation* in which the specifications are expressed in some language, the design is expressed in possibly a different language, and it is to be proven that the design "implements" the specification. There can actually be several stages of such relations; then one has a hierarchy of models, each of which is proved to be an implementation of the one on the next higher level. A structure of this type is obtained by the design method of *stepwise refinement*, in which one starts on an abstract level that allows relatively easy verification, and then proceeds in steps to more and more refined designs which are verified to be specific ways of carrying out the operations on the previous level. Conversely, given a fully specified model, one can try to simplify the model for the purpose of checking a particular property by leaving out detail that is believed to be irrelevant in connection with that property. Typically one then gets a nondeterministic model, in which a precise but complicated description of a part of the system is replaced by some coarse information which nevertheless may still be enough to establish the desired property. This process is sometimes called *abstraction*. Some reduction of complexity may also be achieved by choosing suitable representations of sets of states; in particular it is often useful to work with Boolean expressions which can describe certain sets of states in a compact way. Methods that use such expressions come under the heading of *symbolic model checking*. In addition to model checkers there are also *equivalence checkers* which verify the correctness of a design after modification on the assumption that the original design was correct. Of course the feasibility of various methods to overcome the complexity barrier also depends on the way that programs are formed, and in this context the idea of *modular programming* is important.

A general caveat that should be kept in mind is that formal verification can never give absolute certainty about the functioning of a piece of computer equipment. The reason for this is that verification methods operate on a formal model, which relates to the real world through a formalization step that may contain mistakes. For instance, if a design is formally proved correct but the implementation routine which takes the design to chip circuitry contains an error, then the result may still be faulty. Reportedly this is what happened in the case of the infamous Pentium division bug.

In spite of such incidents, formal verification methods and in particular model checking constitute one of the success stories of computer science, especially in the area of verification of hardware descriptions. All of the major computer firms have developed their own automated verification tools, and commercial tools are finding their way to the marketplace. It is no surprise, therefore, that a large part of the interest of computer scientists who work with hybrid systems goes to verification. Although a general formulation of correctness for hybrid systems should be in terms of language inclusion (and one might discuss which languages are suitable for this purpose), certain properties of interest can be expressed more simply as *reachability* properties, so that the legal trajectories are those in which certain discrete states are *not* reached. This is illustrated in the next subsection.

5.1.2   An audio protocol

Example 5.1.1. A certain protocol for communication between the subsystems of a consumer audio systems was suggested as a benchmark in [24]. The problem is to verify that the protocol is correct in the sense that is ensures that no communication errors occur between the subsystems. Without going into the details of the actual implementation (see [24] for a more elaborate description), the protocol can be described as follows. The purpose of the protocol is to allow the components of a consumer audio system to exchange messages. A certain coding scheme is used which depends on time intervals between events, an "event" being the voltage on a bus interface going from low to high. A basic time interval is chosen and the coding can be described as follows:

- the first event always signifies a 1 (messages always start with a 1);

- if a 0 has last been read, the next event signifies 1 if 4 basic time intervals have passed, 0 if 6 intervals have passed, and 01 if 8 intervals have passed;

- if a 1 has last been read, the next event signifies 0 if 4 intervals have passed, and 01 if 6 intervals have passed;

- if more than 9 intervals pass after a 0 has been read, or more than 7 after a 1 has been read, the message is assumed to have ended.

Due to clock drift and priority scheduling, the timing of events is uncertain and the design specifications call for a 5% tolerance in timing. Clocks are reset with each event.

To describe the protocol using event-flow formulas, it is convenient to model the sender and the receiver separately. Let us begin with the description of the sender. We use two continuous variables, namely clock time denoted by x_s and an offset signal denoted by u_s. There are two discrete state variables A_s and L_s. The variable A_s takes the value 1 if currently a message is being sent and 0 otherwise. The variable L_s denotes the last symbol that has been sent, which is either 0 or 1. There is also a discrete communication variable denoted by S_s representing the signal to be sent to the receiver; this signal is prescribed from outside in a way that is not modeled. Because in the definition of event-flow formulas we have taken as a default that events are not synchronized, we must also introduce a communication variable between sender and receiver which we denote by C. Technically this corresponds to the fact that both sender and receiver are aware of voltage surges on the connecting bus. The operation of the sender can now be described as follows:

$$\text{sender} \ : \ \text{clock\_sender} \ || \ \text{events\_sender} \tag{5.1a}$$

$$\text{clock\_sender} \ : \ \dot{x}_s = 1 + u_s, \quad -0.05 \le u_s \le 0.05 \tag{5.1b}$$

$$\text{events\_sender} \ : \ C = 1, \quad x_s^+ = 0, \ \left| \ \begin{array}{ll} A_s^- = 0, & \text{wakeup\_s} \\[4pt] A_s^+ = 1, & \left| \begin{array}{ll} L_s^- = 0, & \text{event0\_s} \\[2pt] L_s^- = 1, & \text{event1\_s} \end{array} \right. \end{array} \right. \tag{5.1c}$$

$$\text{wakeup\_s} \ : \ S_s = 1, \quad A_s^+ = 1, \quad L_s^+ = 1 \tag{5.1d}$$

$$\text{event0\_s} \ \left| \begin{array}{lll} x_s = 4, & S_s = 1, & L_s^+ = 1 \\ x_s = 6, & S_s = 0, & L_s^+ = 0 \\ x_s = 8, & S_s = 01, & L_s^+ = 1 \\ x_s \ge 9, & A_s^+ = 0 \end{array} \right. \tag{5.1e}$$

$$\text{event1\_s} \ \left| \begin{array}{lll} x_s = 4, & S_s = 0, & L_s^+ = 0 \\ x_s = 6, & S_s = 01, & L_s^+ = 1 \\ x_s \ge 7, & A_s^+ = 0. \end{array} \right. \tag{5.1f}$$

Now we model the receiver. While the sender acts at integer time points as measured on its own clock, the events do not always take place after an integer number of time units on the receiver's clock, as a result of clock inaccuracies. Let us assume that the receiver will round times to the nearest integer in the set $\{4, 6, 8, 9\}$. The operation of the receiver may then be described as follows, notation being similar to the one used for the sender:

$$\text{receiver} \ : \ \text{clock\_receiver} \ || \ \text{events\_receiver} \ || \ \text{timelimit} \tag{5.2a}$$

$$\text{clock\_receiver} \ : \ \dot{x}_r = 1 + u_r, \quad -0.05 \le u_r \le 0.05 \tag{5.2b}$$

$$\text{events\_receiver} : C = 1, \quad x_r^+ = 0, \quad \left|\begin{array}{l} A_r^- = 0, \quad \text{wakeup\_r} \\[2mm] A_r^+ = 1, \quad \left|\begin{array}{ll} L_r^- = 0, & \text{event0\_r} \\[1mm] L_r^- = 1, & \text{event1\_r} \end{array}\right. \end{array}\right. \tag{5.2c}$$

$$\text{wakeup\_r} : S_r = 1, \quad A_r^+ = 1, \quad L_r^+ = 1 \tag{5.2d}$$

$$\text{event0\_r} \quad \left|\begin{array}{lll} x_r < 5, & S_r = 1, & L_r^+ = 1 \\[1mm] 5 \le x_r < 7, & S_r = 0, & L_r^+ = 0 \\[1mm] 7 \le x_r < 9, & S_r = 01, & L_r^+ = 1 \\[1mm] x_r = 9, & A_s^+ = 0 \end{array}\right. \tag{5.2e}$$

$$\text{event1\_r} \quad \left|\begin{array}{lll} x_r < 5, & S_r = 0, & L_r^+ = 0 \\[1mm] 5 \le x_r < 7, & S_r = 01, & L_r^+ = 1 \\[1mm] x_r = 7, & A_r^+ = 0 \end{array}\right. \tag{5.2f}$$

$$\text{timelimit} : L_r = 1, \quad x_r \le 9 \quad | \quad L_r = 0, \quad x_r \le 7. \tag{5.2g}$$

We consider solutions in $NZ/1/C^1/C^0$ of the system as a whole, which is described by

$$\text{system} : \text{sender} \parallel \text{receiver}. \tag{5.3}$$

In particular we are interested in the question whether the discrete external communication traces of sender and receiver are identical for all solutions.

The protocol is said to be *correct* if the string of output symbols is always equal to the string of input symbols. The correctness condition can be described as a (discrete) *reachability condition*: no discrete states (s_1, s_2) with $s_1 \ne s_2$ should be reachable. This means that transitions to such states should never be possible. The verification of this condition requires inspection of the jump conditions and the transition rules, and since these involve continuous dynamics, we have at each discrete state and for each discrete input value a *continuous* reachability problem. Actually in the present case the continuous dynamics is the same at each discrete state, so it suffices to draw a single picture. In Fig. 5.1 the reachable set of continuous states (x_1, x_2) is indicated (shaded) together with the set of points that should be avoided in order to prevent illicit transitions. The reachable set is a cone whose width is determined by the tolerance of the clocks; it is seen that the 5% tolerance is enough (although only barely so) to ensure the correctness of the protocol. An interesting feature of the example is that although as far as the description of the dynamics is concerned there would be no need to distinguish for instance between the discrete states $(1, 1)$ and $(1, 01)$, this distinction does however become important if one wants to formulate correctness requirements. The example also shows that such requirements may take the form of reachability conditions that are ultimately formulated in the continuous state space.

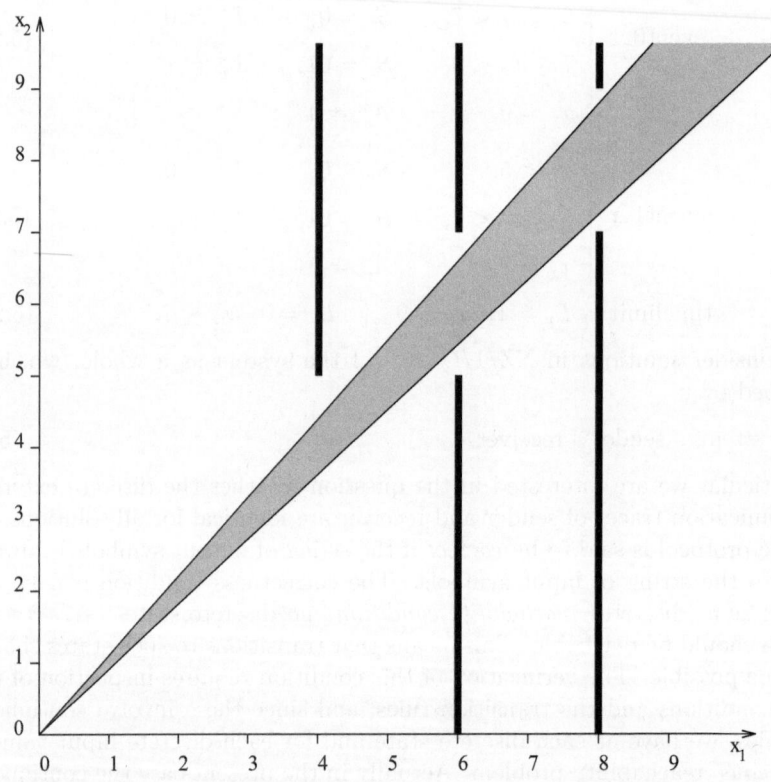

Figure 5.1: Reachable set and avoidance set for audio protocol

5.1.3 Algorithms for verification

Given that verification problems can often be written as reachability problems, the question arises whether it is possible in principle to write an algorithm that will decide, given a hybrid system and an initial condition, whether a certain state is reachable or not. For finite automata, it is easy to construct such an algorithm, so that the discussion in this case would concentrate on the efficiency of algorithms. In the case of continuous systems on the other hand, a simple test is available for the case of linear finite-dimensional time-invariant systems, but there is no general test for nonlinear systems and it would be too much to expect that such a test can be found. So along a scale that begins with finite automata and ends with general hybrid systems, there should be somewhere a point that divides the system types that can be tested for reachability from those for which this is not the case.

It has been shown that for *timed automata* the reachability problem is decidable [1]. Timed automata can be looked at as hybrid systems in which there is only one continuous variable called "time", which satisfies the differential equation $\dot{t} = 1$. The inclusion of this variable makes it possible to express quantitative statements ("after the command `login`, connection to the host computer takes place within three seconds") rather than only qualitative statements such as in standard temporal logic ("after the command `login`, connection to the host computer takes place eventually"). From the standpoint of finite automata this is a major extension, from the standpoint of hybrid systems however timed automata form a rather small class.

It was shown in [1] that the reachability problem is still decidable for hybrid systems that have several clocks running at different rates (so continuous variables x_i satisfying differential equations of the form $\dot{x}_i = c_i$ where the c_i's are constants), as long as all inequality constraints involving these clocks are of the form $x_i \geq k$ or $x_i \leq k$ (so not of the form $x_i \geq x_j$). As soon as one allows two clocks that run at different rates and that may be compared to each other, however, the reachability problem becomes undecidable [1, Thm. 3.2]. So it must be concluded that, for hybrid systems in which there is any serious involvement of continuous variables, one cannot expect to find a general algorithm that will decide reachability.

This being the case, one can still attempt to design algorithms that will *sometimes* terminate, in particular for limited classes of hybrid systems. The tool HyTech [76] has been built at Cornell for the reachability analysis of systems with several clocks under constraints that guarantee that the reachable sets in the continuous state space can always be described by sets of linear inequalities. A symbolic model checker for timed automata is Kronos [41], developed at IMAG in Grenoble.

So what can be said about the verification of hybrid systems that involve a significant amount of (nonlinear) continuous dynamics? The problem itself is certainly of importance; there are many cases in which computer programs interact with the continuous "real" world, and it can literally be a matter of life and death to make sure that the interaction takes place as expected. It should be noted in the first place that tools like HYTECH can still be used for the analysis of hybrid systems that do not strictly satisfy their assumptions, since one can approximate the given system by one that does fall within the scope of the tool. If the approximation is done conservatively (so that for instance the reachable set of the approximating system always contains the reachable set of the given system), one can even obtain strict guarantees. In the case of tools that are based on differential equations with constant right hand sides (so that the reachable sets are indeed polyhedral) one will need to approximate a given function by a piecewise constant function. Depending on the nature of the given function, a reasonable approximation may require a large number of pieces, and both the memory requirements and the computation time of a formal verification tool may soon become prohibitive.

An alternative is to resort to simulation: simply generate a large number of scenarios to see if any faults occur. This methodology can provide at least some test of validity in cases where other methods do not apply. Purely random simulation may take a long time to find potentially hazardous situations; therefore one should use rare-event simulation techniques which should be guided as much as possible by information about situations where problems may be expected. Such information might come for instance from model checking on a coarse approximation of the system. Alternatively one may attempt to look systematically for worst-case situations, by introducing an "adversary" who is manipulating disturbance inputs (or more generally any available non-determinism) with malicious intent. The adversary's behavior may be formally obtained as the solution of an optimization problem which may however be difficult to solve; in general, optimization problems in a hybrid system context have not been studied much yet. If at the same time one wants to design optimal strategies for the system to respond to adverse circumstances one obtains a minimax game problem, which may be even harder to solve. Experience in case studies such as the one in [88] will have to determine the ranges of applicability of each of the possible methodologies or combinations of them.

5.2   Stability

5.2.1   Lyapunov functions and Poincaré mappings

In the study of the stability of nonlinear dynamical systems, Lyapunov functions and Poincaré mappings play a central role. Lyapunov functions can be used to prove the stability of equilibria or the invariance of certain sets; such sets are the level sets of a scalar function on the state space which has monotonous behavior along system trajectories. The idea behind Poincaré mappings is that of the stroboscope. By looking at the continuous states only at certain points in time, one obtains a discrete-time system. By analyzing this discrete-time system, one may obtain information about the stability of certain motions of the continuous-time system (for instance periodic orbits).

For nonlinear systems there are in fact many different notions of stability, ranging from local stability to uniform global asymptotic stability. The notion of stability may be applied to equilibrium points but also more generally to invariant sets. Given a dynamical system, a subset M of the state space is said to be (forward) *invariant* if all trajectories that have their initial point in M are in fact fully contained in M. With this terminology, an equilibrium may simply be defined as an invariant set consisting of one point.

In the hybrid context, we have a state space with a continuous and a discrete part. There is no problem in extending the notion of invariance to this setting. Convergence to a discrete equilibrium point means that the system eventually settles in one particular mode; such a property may be of interest in specific applications. Certainly also the notion of invariance of a certain subset of the discrete state space can be of importance, for instance from a verification point of view. Methods that allow proofs of convergence and invariance are therefore at least as important in hybrid systems as they are in smooth dynamical systems.

First let us review a few definitions that are used in stability theory. Consider a dynamical system defined on some state space X which is equipped with some metric d. The state space X may be a product of a continuous space like $\mathbb{R}^k$ (or more generally a differentiable manifold) and a discrete space S. The set S of possible values of the discrete variables will always be assumed to carry the discrete metric, which means that convergence is the same as eventual equality. If one is interested only in stability of the continuous variables, one may consider the dynamical system defined on the continuous part which is obtained by simply omitting the symbolic parts of trajectories. Now let x_0 be an equilibrium point. The equilibrium is said to be *stable* (or sometimes also *stable in the sense of Lyapunov*, or *marginally stable*) if for every $\varepsilon > 0$ there exists a $\delta > 0$ such that every trajectory that starts at a distance less than δ from x_0 will stay within a distance ε from x_0. In other words, according to this definition x_0 is stable if we can guarantee that trajectories will remain arbitrarily close to x_0 by giving them an initial condition sufficiently close to x_0. There is no implication here that trajectories will converge to x_0. In the following notion of stability we do have such an implication. The equilibrium x_0 is said to be *asymptotically stable* if there exists $\varepsilon > 0$ such that all trajectories with initial conditions at distance less than ε to x_0 converge to x_0. Clearly this is a *local* notion of stability. The equilibrium x_0 is said to be *globally asymptotically stable* if *all* trajectories of the system converge to x_0. There are various related notions such as ultimate boundedness and exponential stability that we shall not discuss here; also, all definitions may be generalized in a straightforward way to the case in which the equilibrium x_0 is replaced by an invariant set.

In the standard application of Lyapunov functions to smooth dynamical systems, the stability of an equilibrium can be concluded if it possible to find a continuous scalar function $V(x)$ that has a minimum at x_0 and that is nonincreasing along trajectories. By imposing additional conditions on the Lyapunov function $V(x)$, one can obtain stronger properties such as asymptotic stability and global asymptotic stability. At first sight it would seem that the condition that $V(x)$ should be nonincreasing along trajectories depends on knowing the solutions. Of course, any method that requires the computation of solutions is uninteresting, firstly because it is usually not possible to compute the solutions explicitly, and secondly because the stability problem becomes trivial once the solutions are known (because one can then see immediately from the computed trajectories whether stability holds). Fortunately, if $V(x)$ is differentiable and the system is given by a differential equation $\dot{x}(t) = f(x(t))$, one can verify that $V(x)$ is nonincreasing along trajectories, without knowing the solutions, by checking that $(\partial V/\partial x)(x)f(x) \leq 0$. In the context of hybrid systems however this simple check is not available, since the system is not given by a single differential equation and there may be jumps in the state trajectories at event times. For this reason the transplantation of Lyapunov theory from the continuous to the hybrid domain is not straightforward.

The most direct generalization of the method of Lyapunov functions to hybrid systems would be the following (see for instance [161]). Suppose there is a continuous function $V(x)$ defined on the continuous state space which has a minimum at an equilibrium point x_0, is nondecreasing along trajectories on interevent intervals, and satisfies $V(x^+) \leq V(x^-)$ whenever x^- and x^+ are connected through a jump rule. Under these conditions, the equilibrium x_0 is stable. With appropriate additional conditions on $V(x)$, stronger conclusions can be drawn; also, analogous statements can be made for invariant sets instead of equilibrium points. As is the case for smooth nonlinear systems, there are no general rules for constructing functions that satisfy the above properties. For hybrid systems in which jumps occur, the condition that $V(x)$ should also be nonincreasing across jumps may introduce an additional complication. An example in which a Lyapunov function is easy to find is the class of unilaterally constrained mechanical systems with Moreau's switching rule for inelastic collisions, where the energy can be taken as such (see Section 4.5).

In some cases, information from the event conditions may make it easier to construct a Lyapunov function. For instance, in the case of the bouncing ball (Subsection 2.2.3) it is natural to take the energy as a Lyapunov function, and it follows immediately from the event conditions that $V(x^+) = e^2 V(x^-)$ at event times, so that stability (even global asymptotic stability) of the equilibrium follows if the restitution coefficient e is less than 1. In between events, the energy is constant and so without the information from the event conditions one could only conclude stability in Lyapunov's sense. In general, stability of an equilibrium point will follow if one can find a scalar function V satisfying suitable growth conditions and having a minimum 0 at x_0, which is nonincreasing on the sequence $\tau_1, \tau_2, \cdots$ of event times and for which an inequality of the form $V(x(t)) \leq h(V(x(\tau_k)))$ $(t \in (\tau_k, \tau_{k+1}))$ can be proven, where h is a nonnegative function satisfying $h(0) = 0$ [161, 127].

Given the extremely wide scope of the class of hybrid dynamical systems, perhaps not too much should be expected from studies of stability that are conducted on this level of generality; for specific classes, it might be easier to come up with verifiable stability criteria. For systems that move between several operating conditions each related to a particular region of the state space it may be a good idea to look for *multiple* Lyapunov functions defined on each of the operating regions. For systems that can be written as feedback systems consisting of a "plant" satisfying passivity conditions and a static feedback the hyperstability theory due to Popov [130] may be useful; here the proofs also ultimately depend on Lyapunov functions, but as a result of the special assumptions it is possible to give specific recipes for the construction of these functions. In situations in which one has design freedom for instance in the choice of a control scheme, a possible approach is to try to ensure stability by choosing some function and making it a Lyapunov function by a suitable choice of the design parameters.

Stability of equilibria is of interest for systems that are supposed to stay close to some operating point. Many systems however operate in a periodic manner, and in that case one is interested in the stability of periodic orbits. An often used instrument in this context is the *Poincaré map* (sometimes also called *return map*). In the case of smooth dynamical systems, the idea of the return map works as follows. Given some periodic orbit that is to be checked for stability, take a point on the orbit and a surface through that point that is transversal to the orbit. Trajectories that start on the surface sufficiently close to the chosen point will intersect the surface again after a time that is approximately equal to the period of the orbit under investigation. The map which takes the initial point of the trajectory to the point where it intersects the chosen surface again is called the Poincaré map or return map. It is mapping defined on a neighborhood of the point that we started with; one could look at the map as defining a discrete-time dynamical system on a space whose dimension is one lower than that of the original state space. It is in general a nonlinear system, but the point on the orbit that we selected is an equilibrium (because of the periodicity of the orbit) and so we can linearize the system around this equilibrium by taking the Jacobian of the return mapping. If the linearized system is stable (i.e. all eigenvalues are inside the unit circle), then we know that the periodic orbit is attracting so that the corresponding periodic regime is stable. Note that the method requires the computation of the Jacobian of the return map, which may be done on the basis of a linearization of the system in a neighborhood of the periodic orbit. The computation may not be very easy, but at least it does not require finding the solutions of the full original nonlinear system.

In the context of hybrid systems, the idea of using a return map may even be more natural than in the smooth context, since any switching surfaces that occur provide natural candidates for the transversal surfaces on which the return map is defined. One has to be careful however since a small perturbation of an initial condition will in general have an effect on event times or might even change the order of events. Especially in cases where the dynamics between events is fairly simple (for instance in piecewise linear systems), the Poincaré map can nevertheless be a very efficient tool.

5.2.2 Time-controlled switching

If a dynamical system is switched between several subsystems, the stability properties of the system as a whole may be quite different from those of the subsystems. This can already be illustrated in switching between two linear systems, as is shown by the following calculation.

Consider a system that follows the dynamics $\dot{x}(t) = A_1 x(t)$ for a period $\frac{1}{2}\varepsilon$, then switches to $\dot{x}(t) = A_2 x(t)$ for again a period $\frac{1}{2}\varepsilon$, then switches back, and so on. An event-flow formula for the system can be written down as follows:

$$
\left|
\begin{array}{l}
\dot{\tau} = 1, \quad \tau \le \frac{1}{2}\varepsilon, \\
\\
\tau^- = \frac{1}{2}\varepsilon, \quad \tau^+ = 0,
\end{array}
\right.
\left|
\begin{array}{ll}
P = 1, & \dot{x} = A_1 x \\
P = 2, & \dot{x} = A_2 x \\
\\
P^- = 1, & P^+ = 2 \\
P^- = 2, & P^+ = 1
\end{array}
\right.
\tag{5.4}
$$

Let us consider a time point t_0 at which the system begins a period in mode 1, with continuous initial state x_0. At time $t_0 + \frac{1}{2}\varepsilon$, the state variable has evolved to

$$
x(t_0 + \tfrac{1}{2}\varepsilon) = \exp(\tfrac{1}{2}\varepsilon A_1)x_0 = x_0 + \frac{\varepsilon}{2}A_1 x_0 + \frac{\varepsilon^2}{8}A_1^2 x_0 + \cdots
$$

(this is a consequence of the general rule $\exp(A) = \sum_{k=0}^{\infty}(1/k!)A^k$). At time $t_0 + \varepsilon$, we get

$$
\begin{aligned}
x(t_0 + \varepsilon) &= (I + \tfrac{\varepsilon}{2}A_2 + \tfrac{\varepsilon^2}{8}A_2^2 + \cdots)(I + \tfrac{\varepsilon}{2}A_1 + \tfrac{\varepsilon^2}{8}A_1^2 + \cdots)x_0 \\
&= (I + \varepsilon[\tfrac{1}{2}A_1 + \tfrac{1}{2}A_2] + \tfrac{\varepsilon^2}{8}[A_1^2 + A_2^2 + 2A_2 A_1] + \cdots)x_0.
\end{aligned}
$$

If we compare the above expression to the power series development for $\exp[\varepsilon(\tfrac{1}{2}A_1 + \tfrac{1}{2}A_2)]$ which is given by

$$
\exp[\varepsilon(\tfrac{1}{2}A_1 + \tfrac{1}{2}A_2)] = I + \varepsilon[\tfrac{1}{2}A_1 + \tfrac{1}{2}A_2] + \frac{\varepsilon^2}{8}[A_1^2 + A_2^2 + A_1 A_2 + A_2 A_1] + \cdots
$$

we see that the constant and the linear term are the same, whereas the quadratic term is off by an amount of $A_1 A_2 - A_2 A_1$ (the "commutator" of A_1 and A_2). So the difference between the solution of the switched system and that of the smooth system $\dot{x} = (\tfrac{1}{2}A_1 + \tfrac{1}{2}A_2)x$ on a time interval of length ε is of the order ε^2. If we let ε tend to zero, then on any fixed time interval the solution of (5.4) will tend to the solution (with the same initial condition) of the "averaged" system

$$
\dot{x}(t) = (\tfrac{1}{2}A_1 + \tfrac{1}{2}A_2)x(t). \tag{5.5}
$$

In particular, the stability properties of the switched system (5.4) will for small ε be determined by the stability properties of the averaged system (5.5), that is, by the location of the eigenvalues of $\frac{1}{2}A_1 + \frac{1}{2}A_2$. Now it is well-known that the eigenvalues are nonlinear functions of the matrix entries and so it may well happen that the matrices A_1 and A_2 are both Hurwitz (all eigenvalues have negative real parts) whereas the matrix $\frac{1}{2}A_1 + \frac{1}{2}A_2$ is unstable, or vice versa. This is illustrated in the following example.

Example 5.2.1. Consider the system (5.4) with

$$A_1 = \begin{bmatrix} -0.5 & 1 \\ 100 & -1 \end{bmatrix}, \quad A_2 = \begin{bmatrix} -1 & -100 \\ -0.5 & -1 \end{bmatrix}. \tag{5.6}$$

Although A_1 and A_2 are both unstable, the matrix $\frac{1}{2}(A_1 + A_2)$ is Hurwitz. Therefore the switched system should be stable if the frequency of switching is sufficiently high. The switching frequency that is minimally needed can be found by computing the eigenvalues of the mapping $\exp(\frac{1}{2}\varepsilon A_1)\exp(\frac{1}{2}\varepsilon A_2)$ as a function of ε; stability is achieved when both eigenvalues are inside the unit circle. For the present case, it turns out that a switching frequency of at least 50 Hz is needed. The plot in Fig. 5.2 shows a trajectory of the system when it is switched at 100 Hz.

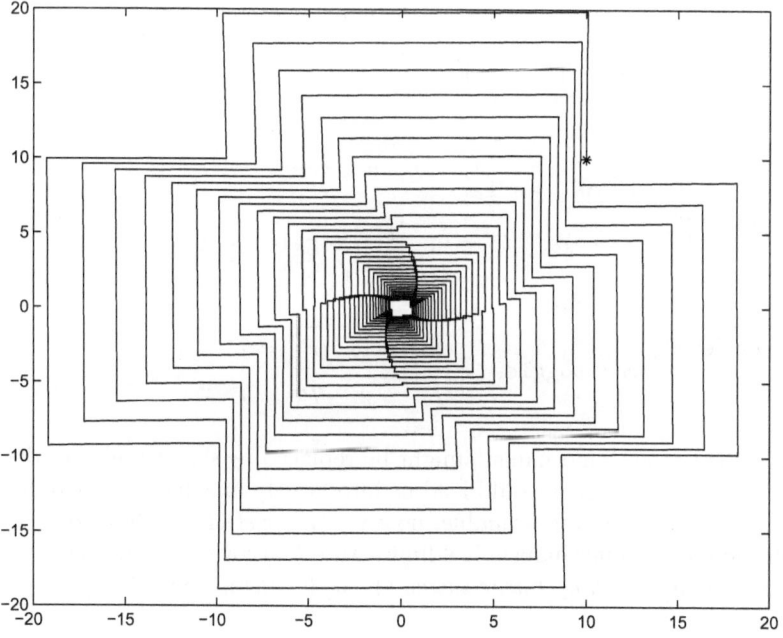

Figure 5.2: Trajectory of the switched system defined by (5.6). Initial point marked by a star. Switching frequency 100 Hz; simulation period 2 units of time

Instead of staying both in mode 1 and in mode 2 for a time interval of length $\frac{1}{2}\varepsilon$, we can also consider the situation in which mode 1 is followed during a time $h\varepsilon$ and mode 2 during a time interval $(1-h)\varepsilon$, where h is a number between 0 and 1. By the same reasoning as above, the behavior of such systems on fixed time intervals is well approximated by that of the system $\dot{x}(t) = A_h x(t)$, where $A_h = hA_1 + (1-h)A_2$. The choice of the parameter h influences the system dynamics and so h might be considered as a control input; the averaging analysis requires, however, that if h varies it should be on a time scale that is much slower than the time scale at which the switching takes place. When mode 1 corresponds to "power on" and mode 2 is "power off", the parameter h is known as the *duty ratio*. In power electronics the use of switches is popular because theoretically it provides a possibility to regulate power without loss of energy. Fast switching is a particular form of this method of regulation; for further discussion see 6.2.2.

Remark 5.2.2. Using a standard trick in differential equations, one may consider the time parameter t as a state variable satisfying the simple differential equation $\dot{t} = 1$. The systems considered in this subsection may then be seen as parallel compositions of two hybrid systems which are coupled via a discrete communication channel. For instance, for the case of equidistant switching instants one may write an event-flow formula as follows:

$$\text{system} = \text{top-level} \parallel \text{low-level} \tag{5.7a}$$

with

$$\text{top-level} \left| \begin{array}{ll} \dot{t} = 1, & t \leq \frac{1}{2}\varepsilon \\ t^- = \frac{1}{2}\varepsilon, & t^+ = 0, \quad S = \texttt{toggle} \end{array} \right. \tag{5.7b}$$

$$\text{low-level} \left| \begin{array}{l} \dot{x} = A_P x \\ S = \texttt{toggle}, \end{array} \right| \left. \begin{array}{ll} P^- = 1, & P^+ = 2 \\ P^- = 2, & P^+ = 1. \end{array} \right. \tag{5.7c}$$

Clearly the top-level mechanism might be replaced by something considerably more complicated, which would lead us into largely unexplored research territory. Stability results are available, however, for systems of the above type in which the top-level mechanism is a finite-state Markov chain and the low-level systems are linear (*jump linear systems*); see for instance [107, 52].

5.2.3   State-controlled switching

In the previous subsection we have considered systems in which the switching is not at all influenced by the continuous state whose stability we are interested in. Let us now take an opposite view and consider systems in which the switching is completely determined by the continuous state. Below we shall consider in particular cases in which the discrete state of the system is determined by the *current* continuous state, so that there is no "discrete memory" or "hysteresis". For such systems the continuous state space can be thought of as being divided into cells which each correspond to a particular discrete state, so that each cell has its own continuous dynamics associated to it. Switching of this type is often referred to as *state-controlled*, with an implicit identification of "state" and "continuous state". Compare also the distinction between internally and externally induced events that was discussed in Chapter 1. We shall concentrate in particular on cases where the trajectories of the continuous state variables are continuous functions of time (no jumps) and the cells all have equal dimension (no motions along lower-dimensional surfaces).

Specifically, let the continuous state space X be divided into cells X_i corresponding to discrete states $i \in I$, and suppose that the dynamics in each mode i is given by the ODE $\dot{x}(t) = f_i(x(t))$. To prove for instance stability of the origin, a possible strategy is to look for a Lyapunov function $V(x)$ which in particular should be such that $\frac{\partial V}{\partial x}(x)f_i(x) \leq 0$ when $x \in X_i$. Finding a Lyapunov function for systems with a single mode is already in most cases a difficult problem; in the multimodal case things can only be expected to be worse in general. Below we shall concentrate attention on situations in which the cells are described by linear inequalities and the dynamics in each cell is linear.

In the single-mode case, the theory of constructing a Lyapunov function for linear systems is classical. Given a linear ODE of the form $\dot{x}(t) = Ax(t)$ with $A \in \mathbb{R}^{n \times n}$, one looks for a quadratic Lyapunov function, that is, one of the form $V(x) = x^T P x$ where P is a positive definite matrix. For such a function, the expression $\frac{\partial V}{\partial x}$ is a quadratic form that we may write as $x^T(A^T P + PA)x$, and so the function $V(x) = x^T P x$ qualifies as a Lyapunov function for the system $\dot{x} = Ax$ if and only if the matrix $A^T P + PA$ is negative definite. To find a Lyapunov function of this form we may solve the linear equation $A^T P + PA = -Q$ where Q is some given positive definite matrix (take for instance $Q = I$) and where P is the unknown. A standard result in stability theory asserts that, whenever the matrix A is Hurwitz, this equation has a unique solution which is positive definite. So, asymptotic stability of linear systems can always be proved by a quadratic Lyapunov function, and such a Lyapunov function can be found by solving a linear system of $\frac{1}{2}n(n+1)$ equations in $\frac{1}{2}n(n+1)$ unknowns. The stability analysis of linear systems is therefore computationally quite feasible, in sharp contrast to the situation for general nonlinear systems.

In the multimodal case it is a natural idea to look for piecewise quadratic Lyapunov functions when the dynamics in all modes are linear. In general, a dynamical system defined on $\mathbb{R}^n$ is said to be *quadratically stable* with respect to the origin if there exists a positive definite matrix P and a scalar $\varepsilon > 0$ such that $d(x^T P x)/dt \leq -\varepsilon x^T x$ along trajectories of the system. We shall concentrate here on quadratic stability with respect to the origin, although there are of course other types of stability, and stability of periodic orbits is in many applications perhaps even more interesting than stability of equilibria.

Consider now a piecewise linear system; specifically, let the dynamics in mode i be given by $\dot{x} = A_i x$. The strongest case of stability occurs when all the modes have a common Lyapunov function. If the Lyapunov function is quadratic, this means that there exists a symmetric positive definite matrix P such that

$$A_i^T P + P A_i < 0 \tag{5.8}$$

for all i. In this case the stability does not even depend on the switching scheme. Note that the common Lyapunov function may be found, if it exists, by solving a system of linear matrix inequalities (LMIs).

The condition (5.8) requires in particular that all the constituent matrices A_i are Hurwitz. One can easily find examples, however, in which state-controlled switching of unstable systems leads to an asymptotically stable system. Consider for instance the two linear systems given by the matrices

$$A_1 = \begin{bmatrix} 0.1 & -5 \\ 1 & 0.1 \end{bmatrix}, \quad A_2 = \begin{bmatrix} 0.1 & -1 \\ 5 & 0.1 \end{bmatrix}. \tag{5.9}$$

Figure 5.3 shows trajectories of the systems $\dot{x} = A_1 x$ and $\dot{x} = A_2 x$ starting from $(1,0)$ and $(0,1)$ respectively. Clearly both systems are unstable. However, if we form the switched system which follows the law $\dot{x} = A_1 x$ on the first and third quadrant and the law $\dot{x} = A_2 x$ on the second and fourth quadrant, then the result is asymptotically stable as can be verified in this case by computing the trajectories of the switched system explicitly. By reversing time in this example one gets an example of an unstable system that is obtained from switching between two asymptotically stable systems.

For a state-controlled switching scheme, the condition (5.8) can be replaced by the weaker condition

$$\text{for all } i \in I: \quad x^T(A_i^T P + P A_i)x < 0 \quad \text{for all } x \in X_i, \, x \neq 0. \tag{5.10}$$

which still implies asymptotic stability of the system subject to state-controlled switching. The above condition is even sufficient for asymptotic stability in case a sliding mode occurs along the boundary of two regions X_i and X_j, since the conditions $x^T(A_i^T P + P A_i)x < 0$ and $x^T(A_j^T P + P A_j)x < 0$ imply $x^T(A^T P + P A)x < 0$ for all convex combinations A of A_i and A_j. A concrete way of using the criterion (5.10) is to look for symmetric matrices S_i such that $x^T S_i x \geq 0$ for $x \in X_i$. The existence of a positive definite matrix P such that for all i

$$A_i^T P + P A_i + S_i < 0 \tag{5.11}$$

is then sufficient for asymptotic stability. This is the so-called S-procedure (see e.g. [25]).

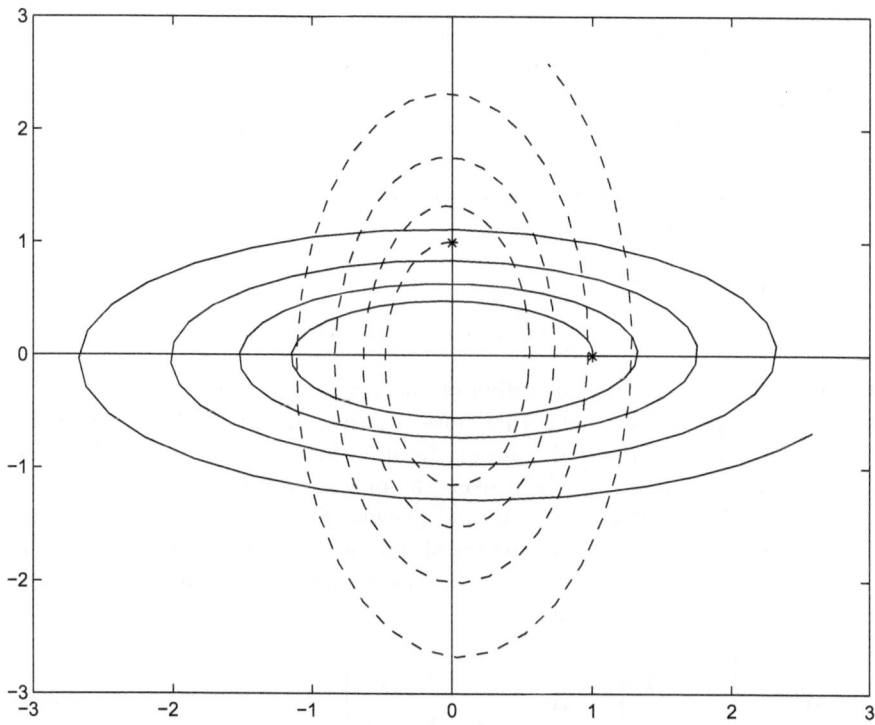

Figure 5.3: Trajectories of the systems (5.9); initial points marked by stars

To further weaken the conditions for stability, one may look at Lyapunov functions that are piecewise quadratic rather than quadratic. If a region-controlled system of the above type is considered, then a natural idea is to look for a function of the form $x^T P_i x$ for $x \in X_i$, where one should have $x^T P_i x = x^T P_j x$ when x is in a common boundary of regions X_i and X_j, in order to make the definition consistent and to make the standard stability arguments work. As a particular case, consider a system of the form

$$\dot{x} = A_1 x \text{ if } Cx \geq 0, \quad \dot{x} = A_2 x \text{ if } Cx \leq 0 \tag{5.12}$$

where C is a linear mapping from $\mathbb{R}^n$ to $\mathbb{R}$ and where it is assumed that no sliding mode occurs, so that we do not necessarily need to assume that $A_1 x = A_2 x$ when $Cx = 0$. Stability will be guaranteed if we can find symmetric positive definite matrices P_1 and P_2 such that $A_i^T P_i + P_i A_i < 0$ for $i = 1, 2$, and $x^T P_1 x = x^T P_2 x$ when $Cx = 0$. The latter condition is fulfilled when P_1 and P_2 are given by $P_1 = P$, $P_2 = P + \alpha C^T C$ for some real α. So a sufficient condition for stability of the system (5.12) is the existence of a matrix P and a scalar α such that

$$P = P^T > 0, \quad A_1^T P + P A_1 \leq 0, \quad P + \alpha C^T C > 0,$$
$$A_2^T (P + \alpha C^T C) + (P + \alpha C^T C) A_2 \leq 0. \tag{5.13}$$

Note that (5.13) constitutes a system of linear matrix inequalities. The conditions above imply in particular that both matrices and A_1 and A_2 are stable, so that the suggested method can only be used for systems composed of stable subsystems. There is no help from the observation that it is sufficient if $x^T(A_1^T P + P A_1)x \leq 0$ for $Cx \geq 0$, and likewise for the other mode, because the condition is invariant under change of sign of x. It can be seen from simple examples that a system of the form (5.12) may well be asymptotically stable even if A_1 and A_2 are not both stable; take for instance

$$A_1 = \begin{bmatrix} 1 & 1 \\ -1 & 1 \end{bmatrix}, \quad A_2 = \begin{bmatrix} -2 & 1 \\ -1 & -2 \end{bmatrix} \tag{5.14}$$

together with any nonzero functional C.

An even further reduction of conservatism can be obtained by dropping the requirement that the Lyapunov function should be continuous; indeed, there is no problem with discontinuities as long as jumps that take place along trajectories are downward. Again, this observation is of no help when the considered Lyapunov functions are piecewise quadratic and the cells are halfspaces. In other cases, however, one may write down the inequalities that should be satisfied by a piecewise quadratic but not necessarily continuous Lyapunov function and attempt to solve the resulting system by LMI techniques.

Remark 5.2.3. In the above we have assumed in a number of places that no sliding modes occur. The presence of motions along lower-dimensional manifolds may definitely affect the stability analysis. As shown in the following simple example, a sliding mode can be unstable even when it is composed from two stable dynamics.

Example 5.2.4. Consider the systems $\dot{x} = A_1 x$ and $\dot{x} = A_2 x$ with

$$
A_1 = \begin{bmatrix} 1 & 2 \\ -2 & -2 \end{bmatrix}, \quad A_2 = \begin{bmatrix} 1 & -2 \\ 2 & -2 \end{bmatrix} \tag{5.15}
$$

and suppose that mode 1 is valid for $x_2 < 0$ and mode 2 for $x_2 > 0$. It is easily verified that both modes would be asymptotically stable if they would be valid on the whole plane. On the line $x_2 = 0$ there is a sliding mode which is obtained by averaging the two dynamics, and which results in an unstable motion $\dot{x}_1 = x_1$.

5.3 Chaotic phenomena

As is well-known, smooth dynamical systems of dimension three and higher may give rise to complicated behavior. Although there is no universally accepted definition of "chaos", the term is usually associated to phenomena such as sensitive dependence on initial conditions and the presence of attractors containing several dense orbits. In the literature, several examples are known of situations in which a few simple smooth systems (for instance linear systems) give rise to chaotic phenomena when they are combined into a hybrid system by some switching rule. Here we briefly discuss three of these examples.

Example 5.3.1 (A switched arrival system). The following system was described in [36]. Consider N buffers from which work is removed at rates ρ_i $(i = 1, \ldots, N)$. A server delivers work to one of the buffers at rate 1. It is assumed that the total amount of work in the system is constant, so that $\sum_{i=1}^{N} \rho_i = 1$. As soon as one of the buffers becomes empty, the server switches to that buffer. The system has N continuous variables x_i which denote the amounts of work in each of the buffers, and one discrete variable j which denotes the buffer currently being served. The dynamics of the system is given by the EFF

$$
\{\dot{x} = e_j - \rho\} \ \| \ |_{i \in \overline{N}} \{e_i^T x^- = 0, \ j^+ = i\} \tag{5.16}
$$

where x is the vector of buffer contents, ρ is the constant vector of extraction rates, e_i denotes the ith unit vector, and $\overline{N}$ is the set $\{1, \ldots, N\}$. Note that the system is not deadlock-free; if two buffers empty at the same time, there is no way to continue. However it is shown in [36] that for almost all initial conditions this never happens.

For $N = 2$, it is easy to verify that the switched arrival system has a simple periodic behavior. However already for $N = 3$ most trajectories have a highly irregular behavior, and it can be shown that all periodic trajectories are unstable. There is some regularity in a statistical sense, however. The main result of [36] states that there is one probability distribution which is *stationary* in the sense that if the initial condition with one of the buffers empty is drawn from this distribution, then the distribution of the state at the next event time is the same. This distribution gives certain probabilities (depending on the rates ρ_1, ρ_2 and ρ_3) to each of the three buffers being empty, and within these cases assigns a uniform distribution over the contents of the other two buffers, subject, of course, to the requirement that the sum of the contents of the buffers should be equal to the constant amount of work that is always present in the system. It is moreover shown that this distribution is *statistically stable*, which means that if we start with an initial condition that is distributed according to some arbitrary distribution, then the distribution of the state (taken at event times) will converge to the stationary distribution. Finally the *ergodicity* property holds. Ergodicity means that for almost all initial conditions "time averages are the same as sample averages", that is to say that the value of the average over time of any measurable function of the state at event times can be computed as the expected value of that function with respect to the stationary distribution. By this result, it can be computed that for almost all inital conditions the average interevent time for the trajectory starting from that initial condition is equal to $\frac{1}{4}(\rho_1\rho_2 + \rho_2\rho_3 + \rho_1\rho_3)^{-1}$.

Example 5.3.2 (A vibrating system with a one-sided spring). In [74] a mechanical system is described that consists of a vibrating beam and a one-sided spring. The spring is placed in the middle of the beam in such a way that it becomes active or inactive when the midpoint of the beam passes through its equilibrium position. The system is subject to a periodical excitation from a rotating mass unbalance attached to the beam. The model proposed in [74] has three degrees of freedom. The equations of motion are given by

$$M\ddot{q} + B\dot{q} + (K + H)q = Gf \tag{5.17}$$

if the displacement of the midpoint of the beam is positive, and by

$$M\ddot{q} + B\dot{q} + Kq = Gf \tag{5.18}$$

if the displacement is negative. Here, q is the three-dimensional configuration variable, M is the mass matrix, B is the damping matrix, K is the stiffness matrix when the spring is not active, H is the additional stiffness due to the spring, f is the excitation force, and G is an input vector.

The main emphasis in [74] is on control design and no extensive study of chaotic phenomena is undertaken. Still the analysis shows at least that there are several unstable periodic orbits. In the cited paper the aim is to add a controller to the system which will stabilize one of these orbits. Because the control takes place around an orbit that is already periodic by itself (although unstable), one may expect that a relatively small control effort will be needed for stabilization once the desired periodic regime has been achieved. This idea of using only a small control input by making use of the intrinsic nonlinear properties of the given system is one of the leading themes in the research literature on "control of chaos".

Example 5.3.3 (The double scroll circuit). A well-known example of a simple electrical network that shows chaotic behavior is Chua's circuit, described for instance in [38]. The circuit consists of two capacitors, one inductor, one linear resistor and one (active) nonlinear resistor. The equations may be given in the form of a feedback system with a static piecewise linear element in the feedback loop, as follows:

$$
\begin{aligned}
\dot{x}_1 &= \alpha(x_1 - u) \\
\dot{x}_2 &= x_1 - x_2 + x_3 \\
\dot{x}_3 &= -\beta x_2 \\
y &= x_1
\end{aligned}
\tag{5.19}
$$

and

$$
\left|
\begin{aligned}
&y \le -1, \quad u = m_1 y - (m_0 - m_1) \\
&-1 \le y \le 1, \quad u = m_0 y \\
&y \ge 1, \quad u = m_1 y + (m_0 - m_1).
\end{aligned}
\right.
\tag{5.20}
$$

Note that the function defined in (5.20) is continuous, so that the above system may also be looked at as a three-dimensional dynamical system defined by a piecewise linear continuous vector field. The symbols α, β, m_0 and m_1 denote parameters. If one studies the system's behavior in terms of these parameters one finds a rich variety of motions. For some parameter values, trajectories occur that look like two connected scrolls, and for this reason the system above is also known as the "double scroll system". The presence of chaos (in one of the possible definitions) is shown in [38] for the values $m_0 = -1/7$, $m_1 = 2/7$, $\alpha = 7$, and β near 8.6.

The examples above present just a few instances of chaos in nonsmooth dynamical systems; many more examples can be found in the literature. We have chosen examples that exhibit switching between different modes. The earliest examples of chaos in nonsmooth systems are related to what might be called *impacting systems* in which there are state-generated events but no changes of mode. The standard example is the one of a ball bouncing on a vibrating table, which is discussed in the well known textbook [59]. In contexts of this type it is natural to look for parameter changes that will lead from a situation in which there are no impacts to a situation in which one does have (low-velocity) impacts. The associated bifurcations are called *grazing bifurcations* or *border collision bifurcations*; see for instance [32]. Impacting systems can be seen as limiting cases of situations in which a contact regime exists for a very short time period. In many mechanical applications it is reasonable to consider systems that are subject to different regimes for periods of comparable length. The one-sided spring mentioned above is an example of this type. Chaotic behavior has also been shown to occur for instance in periodically excited mechanical systems subject to Coulomb friction, and also in this case one typically sees different regimes that coexist on the same time scale; see for instance [131]. In an electrical circuit context, the occurrence of chaotic phenomena in systems with diodes, and in particular DC-DC converters, is well documented; see for instance [42, 47]. In this context one typically has linear dynamics in each mode. In particular it is noted from these examples that linear complementarity systems, which were extensively discussed in the previous chapter, may exhibit chaotic behavior.

5.4   Notes and references for Chapter 5

The issue of formal verification has been extensively discussed in computer science. Various methodologies exist; see for instance [91, 110]. In the particular context of hybrid systems, many papers related to verification can be found in the workshop proceedings [58, 5, 3, 100, 6, 137, 153]; see also [87]. A classical text on stability is the book by Hahn [60]; for the special case of linear systems see also Gantmacher [56]. A switched linear system can sometimes be looked at as a smooth system with a nonsmooth feedback. The study of such systems has a long history; see for instance Popov's book [130]. In the control context, the issue of stability is closely related to stabilization. This topic will be discussed in the following chapter. In Subsection 5.2.3 we have partly followed the paper [83] by Johansson and Rantzer, which gives a nice survey of quadratic stability for hybrid systems. Other references of interest in this connection include [126, 127]. The literature on chaos is of course enormous; see for instance [59, 123, 134] for entries into the literature. A specific list of references on the analysis of nonsmooth dynamical systems can be found at a useful website maintained at the University of Cologne (www.mi.uni-koeln.de/mi/Forschung/Kuepper/Forschung1.htm).

Chapter 6

Hybrid control design

In this chapter we indicate some directions in hybrid control design. Apart from the introduction of a few general concepts, this will be mainly done by the treatment of some illustrative examples.

The area of hybrid control design is very diverse. The challenging topic of controlling *general* hybrid systems is a wide open problem, and, following up on our discussion in Chapter 1 on the modelling of hybrid systems, it is not to be expected that a powerful (and numerically tractable) theory can be developed for general hybrid systems. Particular classes of hybrid systems for which control problems have been addressed include, among others, batch control processes, power converters and motion control systems, as well as extensions of reactive program synthesis to timed automata. Furthermore, in some situations it is feasible to *abstract* the hybrid system to a discrete-event system (or to an automaton or Petri-net), in which case recourse can be taken to discrete-event control theory. Since our aim is to expose some general methods, we will not elaborate on these more specific areas, but instead indicate in Section 6.1 some general methods of hybrid control design based on game theory and viability theory (or, in control theoretic parlance, the theory of *controlled invariance*). In particular, we discuss the synthesis of controllers enforcing *reachability* specifications.

The main emphasis in this chapter will be on the design of *hybrid controllers* for *continuous-time systems* as ordinarily considered in control theory. From a control perspective this subject has classical roots, and we will actually reformulate some classical notions into the modern paradigm of hybrid systems. In particular, in Section 6.2 we introduce some general terminology on switching control, and we give relations with the classical notions of pulse width modulation and sliding mode control. Furthermore, in continuation of the discussion of quadratic stability in Subsection 5.2.3, we discuss the quadratic *stabilization* of multi-modal linear systems. In Section 6.3 we consider the specific topic of stabilizing continuous-time systems by *switching control schemes*. This, partly classical, area has regained new interest by the discovery that there are large classes of *nonlinear* continuous time systems that are controllable and can be stabilized by switching control schemes, but nevertheless cannot be stabilized by continuous state feedback. Also we will indicate that for controlling certain physical systems, switching control schemes may be an attractive option which allows for a clear physical interpretation.

6.1   Safety and guarantee properties

Let us consider a hybrid system, either given (see Chapter 1) as a (generalized) hybrid automaton (Definitions 1.2.3 and 1.2.5), as a hybrid behavior (Definition 1.2.6), or described by event-flow formulas (Subsection 1.2.6). Consider the total behavior $\mathcal{B}$ of the hybrid system, e.g., in the (generalized) hybrid automaton model, the set of *trajectories* χ of the hybrid system given by functions $P : \tau_{\mathcal{E}} \to L$, $x : \tau_{\mathcal{E}} \to X$, $S : \mathcal{E} \to A$, $w : \tau_{\mathcal{E}} \to W$, for time event sets $\mathcal{E}$, and the corresponding time evolutions $\tau_{\mathcal{E}}$. A *property*, $\mathcal{P}$, of the hybrid system is a map

$$\mathcal{P} : \mathcal{B} \to \{\text{TRUE, FALSE}\}.$$

A trajectory χ satisfies property $\mathcal{P}$ if $\mathcal{P}(\chi) = \text{TRUE}$, and the hybrid system satisfies property $\mathcal{P}$ if $\mathcal{P}(\chi) = \text{TRUE}$ for *all* trajectories $\chi \in \mathcal{B}$.

As already indicated in Chapter 5 in the context of verification, although a general formulation of *correctness* of a hybrid system should be in terms of language inclusion, certain properties of interest of a hybrid system can be often expressed more simply as *reachability* properties of the discrete and continuous state variables, so that legal trajectories are those in which certain states are *not* reached. Thus, let as above L denote the discrete part of the state space and X the continuous part of the state space of the hybrid system. Given a set $F \subset (L \times X)$ we define a *safety property*, denoted by $\Box F$, by

$$\Box F(\chi) = \begin{cases} \text{TRUE if for all } t \in \tau_{\mathcal{E}}, \quad (P(t), x(t)) \in F \\ \\ \text{FALSE otherwise} \end{cases}$$

Also, we can define a *guarantee property*, denoted by $\Diamond F$, by

$$\Diamond F(\chi) = \begin{cases} \text{TRUE if there exists } t \in \tau_{\mathcal{E}} \text{ such that } (P(t), x(t)) \in F \\ \\ \text{FALSE otherwise} \end{cases}$$

(The notations $\Box$ and $\Diamond$ originate from temporal logic, cf. [104].) These two classes of properties are dual in the sense that for all subsets $F \subset L \times X$ and all trajectories χ

$$\Diamond F(\chi) \iff \neg \Box F^c(\chi)$$

where F^c denotes the complement of the set F. Thus, in principle, we may only concentrate on safety properties.

Remark 6.1.1. Of course, one can still define other properties given a subset $F \subset L \times X$. For example, one may require that χ will enter the subset F

at some finite time instant, and from then on will remain in F. This can be seen as a combination of a guarantee and a safety property. Alternatively, one may define a property by requiring that χ enters the set F at infinitely many time instants. Also this property can be regarded as a combination of guarantee and safety (after any finite time instant we are sure that χ will visit F). From a computational point of view these "combined" properties can often be studied by combining the algorithms for safety and guarantee properties, see (especially for the finite automaton case) [102].

6.1.1 Safety and controlled invariance

For *finite automata* (Definition 1.2.2) the study of safety properties, as well as of guarantee properties, is relatively easy. In particular, let us consider automata given by the following generalization of input-output automata (1.3):

$$
\begin{aligned}
l^{\sharp} &\in \delta(l, i) \\
o &= \eta(l, i)
\end{aligned}
\tag{6.1}
$$

where $\delta : L \times I \to 2^L$ defines the possible state successors $l^{\sharp}$ to a state l given the control input $i \in I$. This models automata where the transitions are only *partially* controlled, or *non-deterministic* automata.

Let $F \subset L$ be a given set. Recall that a trajectory χ has property $\Box F$ if its discrete state trajectory $P(t)$ satisfies $P(t) \in F$ for all time instants t. This motivates the following *invariance* definition. First, for any initial state l_0 at time 0, and a control sequence $i_0, i_1, i_2, \ldots$ at the time instants $0, 1, 2, \ldots$, let us denote the resulting state trajectories of (6.1) by $l_0, l_1, l_2, l_3, \ldots$, where $l_1 \in \delta(l_0, i_0)$, $l_2 \in \delta(l_1, i_1)$, $l_3 \in \delta(l_2, i_2)$, and so on. A subset $V \subset L$ is called *controlled invariant* for (6.1) if for all $l_0 \in V$ there exists a control sequence $i_0, i_1, i_2, \ldots$ such that any resulting state trajectory $l_0, l_1, l_2, l_3, \ldots$ of (6.1) remains in V, that is, $l_j \in V$ for $j = 0, 1, 2, \ldots$.

Define for any subset $V \subset L$ its *controllable predecessor* $\mathrm{con}(V)$ by

$$
\mathrm{con}(V) = \{l \in L \mid \exists i \in I \text{ such that } \delta(l, i) \subset V\}.
\tag{6.2}
$$

It follows that V is controlled invariant if and only if

$$
V \subset \mathrm{con}(V).
\tag{6.3}
$$

In order to compute the set of discrete states $l \in F$ for which there exists a control action $i_0, i_1, i_2, \ldots$ such that the state will remain in F for *all* future discrete times $0, 1, 2, \ldots$ we thus have to compute the *maximal* controlled invariant subset contained in F. This can be done via the following algorithm.

$$
\begin{aligned}
V^0 &= F \\
V^{j-1} &= F \cap \mathrm{con}(V^j), \quad \text{for} \quad j = 0, -1, -2, \ldots, \quad \text{until} \\
V^{j-1} &= V^j.
\end{aligned}
\tag{6.4}
$$

Clearly, the sequence of subsets $V^0, V^{-1}, V^{-2}, \ldots$ is non-increasing, and so, by finiteness of F, the algorithm converges in at most $|F|$ steps. At each step of the algorithm, the set V^j contains the discrete states for which there exists a sequence of control actions $i_j, i_{j+1}, i_{j+2}, \ldots, i_{-1}$ which will ensure that the system remains in F for at least $-j$ steps. It is easy to see that the subset obtained in the final step of the algorithm, that is

$$V^*(F) := V^{j-1} = V^j$$

is the *maximal controlled invariant* subset contained in F. Indeed, $V^*(F)$ is clearly a controlled invariant subset of F, since it satisfies by construction

$$V^*(F) = F \cap \operatorname{con}(V^*(F))$$

and therefore $V^*(F)$ satisfies (6.3), as well as $V^*(F) \subset F$. Furthermore, it can be shown inductively that any controlled invariant subset $V \subset F$ satisfies $V \subset V^j$ for $j = 0, -1, -2, \ldots$, and hence $V \subset V^*(F)$, proving maximality of $V^*(F)$.

Hence for every state $l \in V^*(F)$ there exists a control action such that the resulting trajectory satisfies property $\Box F$, and these are precisely all the states for which this is possible.

Note that algorithm (6.4) is constructive also with regard to the required control action: in the process of calculating V^{j-1} we compute for every $l \in V^{j-1}$ a control value i such $\delta(l, i) \subset V^j$. So, at the end of the algorithm we have computed for every $l \in V^*(F)$ a sequence of control values which keeps the trajectories emanating from l in F (in fact, in $V^*(F)$). Furthermore, *outside* $V^*(F)$ we still have complete freedom over the control action.

Using the duality between safety properties and guarantee properties, we can also immediately derive an algorithm for checking the property $\Diamond F$:

$$
\begin{aligned}
S^0 &= F \\
S^{j-1} &= F \cup \operatorname{con}(S^j), \quad \text{for} \quad j = 0, -1, -2, \ldots, \quad \text{until} \\
S^{j-1} &= S^j.
\end{aligned}
\tag{6.5}
$$

Here S^j contains the discrete states from which a visit to F can be enforced after at most $-j$ steps. Furthermore, $S^0, S^{-1}, S^{-2}, \ldots$ is a non-decreasing sequence, and the limiting set

$$S^*(F) := S^{j-1} = S^j$$

defines the set of locations for which there exists a control action such that the resulting trajectory of the finite automaton satisfies property $\Diamond F$.

As noted in Chapter 1 the two most "opposite" examples of hybrid systems are on the one hand finite automata (Definition 1.2.2) and continuous-time systems (Definition 1.2.1). It turns out, roughly speaking, that also for continuous-time systems a safety property $\Box F$ can be checked in a similar way as above, provided the subset F of the continuous state space X is a *submanifold* of X. Indeed, let us consider for concreteness a continuous-time input-state-output system

$$
\begin{aligned}
\dot{x} &= f(x) + g(x)u + p(x)d \\
y &= h(x)
\end{aligned}
\tag{6.6}
$$

where, compared with (1.2), we have split the input variables into a set of *control* variables $u \in \mathbb{R}^m$ and a set of *disturbance* variables $d \in \mathbb{R}^l$. The interpretation is similar to (6.1): part of the input variables can be completely controlled, while the remaining part is totally uncontrollable. Furthermore, as compared with (1.2) we have assumed for simplicity of exposition that the dependence on the input variables u and d is of an *affine* nature via the matrices $g(x)$ and $p(x)$ (see for the general case e.g. [121], Chapter 13).

Now, let F be a given submanifold of the state space X. Denote the *tangent space* of X at a point $x \in X$ by $T_x X$, and the tangent space of any submanifold $N \subset X$ at a point $x \in N$ by $T_x N$. Furthermore, define $G(x)$ as the subspace of $T_x X$ spanned by the image of the matrix $g(x)$, and similarly $P(x)$ as the subspace of $T_x X$ spanned by the image of the matrix $p(x)$.

Consider the algorithm

$$
\begin{aligned}
N^0 &= F \\
N^{j-1} &= \{x \in N^j \mid f(x) + P(x) \subset T_x N^j + G(x)\} \\
&\quad \text{for} \quad j = 0, -1, -2, \ldots, \quad \text{until} \\
N^{j-1} &= N^j
\end{aligned}
\tag{6.7}
$$

where we *assume* that the subsets $N^{j-1} \subset X$ produced in this way are all submanifolds of X (see for some technical conditions ensuring this property, as well as for computing N^{j-1} in local coordinates, [121], Chapters 11 and 13). It is immediately seen that the sequence of submanifolds N^j for $j = 0, -1, -2, \ldots$ is *non-increasing*, that is

$$
N^0 \supset N^{-1} \supset N^{-2} \supset \cdots
$$

and so by finite-dimensionality of F the algorithm will converge in at most $\dim F$ steps to a submanifold

$$
N^*(F) := N^{j-1} = N^j
$$

satisfying the property

$$
f(x) + P(x) \subset T_x(N^*(F)) + G(x), \quad \text{for all } x \in N^*(F)
\tag{6.8}
$$

Also in this case (see e.g. [121]) it can be shown that $N^*(F)$ is the maximal submanifold with this property. For this reason $N^*(F)$ is called the *maximal controlled invariant submanifold* contained in F.

Under a constant rank assumption (see again [121], Chapters 11 and 13, for details) it can be shown that (6.8) implies the existence of a (smooth) state feedback $u = \alpha(x)$ such that the closed-loop system

$$\dot{x} = f(x) + g(x)\alpha(x) + p(x)d \tag{6.9}$$

leaves the submanifold $N^*(F)$ invariant, for any disturbance function $d(\cdot)$. Hence, the trajectories of the closed-loop system (6.9) starting in $N^*(F)$ satisfy property $\square F$.

Algorithm 6.7 can be regarded as an *infinitesimal* version of Algorithm 6.4; it replaces the transitions in the finite automaton by conditions on the velocity vector $\dot{x}$ and works on the tangent space of the state space manifold X. Note also that the finiteness property of F in Algorithm 6.4 has been replaced in Algorithm 6.7 by the finite-*dimensionality* of the submanifold F.

Algorithm 6.7 works well *provided* the set F (as well as all the subsequent subsets $N^{-1}, N^{-2}, N^{-3}, \ldots$) is a submanifold, but breaks down for more general subsets. A fortiori for general hybrid systems it is not clear how to extend the algorithm.

6.1.2   Safety and dynamic game theory

Another approach to checking safety properties of the form $\square F$ is based on dynamic *game theory*. Indeed, consider a continuous-time system

$$\dot{x} = f(x, u, d) \tag{6.10}$$

where, as before in (6.6), $u \in U$ are the control inputs (corresponding to the first player) and $d \in D$ are the disturbance inputs (corresponding to the second player who is known as the *adversary*). However, we allow U and D to be *arbitrary* sets, not necessarily $\mathbb{R}^m$ and $\mathbb{R}^l$, implying that (part of) the control and disturbance inputs may be *discrete*.

Furthermore, contrary to the submanifolds F considered before, we assume that the safety subset F is given by *inequality constraints*, that is

$$F = \{x \in X \mid k(x) \geq 0\} \tag{6.11}$$

where $k : X \to \mathbb{R}$ is a differentiable function with $\frac{\partial k}{\partial x}(x) \neq 0$ on the boundary

$$\partial K = \{x \in X \mid k(x) = 0\}$$

Let now $t \leq 0$, and consider the *cost function*

$$J : X \times \mathcal{U} \times \mathcal{D} \times \mathbb{R}^- \to \mathbb{R} \tag{6.12}$$

where $\mathcal{U}$ and $\mathcal{D}$ denote the admissible control, respectively, disturbance functions, with the end condition

$$J(x, u(\cdot), d(\cdot), t) = k(x(0)). \tag{6.13}$$

This function may be interpreted as the cost associated with a trajectory of (6.10) starting at x at time $t \leq 0$ according to the input functions $u(\cdot)$ and $d(\cdot)$, and ending at time $t = 0$ at the final state $x(0)$. Furthermore, define the *value function*

$$J^*(x, t) = \max_{u \in \mathcal{U}} \min_{d \in \mathcal{D}} J(x, u, d, t). \tag{6.14}$$

Then the set

$$\{x \in X \mid \min_{t' \in [t,0]} J^*(x, t') \geq 0\} \tag{6.15}$$

contains all the states for which the system can be forced by the control u to remain in F for at least $|t|$ time units, irrespective of the disturbance function d.

The value function J^* can be computed, in principle, by standard techniques from dynamic game theory (see e.g. [14]). Define the (pre-)*Hamiltonian* of the system (6.10) by

$$H(x, p, u, d) = p^T f(x, u, d) \tag{6.16}$$

where p is a vector in $\mathbb{R}^n$ called the *costate*. The optimal Hamiltonian is defined by

$$H^*(x, p) = \max_{u \in U} \min_{d \in D} H(x, p, u, d) \tag{6.17}$$

If J^* is a differentiable function of x, t, then J^* is a solution of the *Hamilton-Jacobi* equation

$$-\frac{\partial J^*(x, t)}{\partial t} = H^*(x, \frac{\partial J^*(x, t)}{\partial x}) \tag{6.18a}$$

$$J^*(x, 0) = k(x) \tag{6.18b}$$

It follows that there exists a control action such that the trajectory emanating from a state x has property $\Box F$ if and only if (compare with (6.15))

$$J^*(x, t') \geq 0 \quad \text{for all } t' \in (-\infty, 0]. \tag{6.19}$$

In [99] it is indicated how to simplify this condition by considering a modified Hamilton-Jacobi equation. Indeed, to prevent states from being relabeled as safe once they have been labeled as unsafe ($J^*(x, t)$ being negative for some time t), one may replace (6.18) by

$$-\frac{\partial J^\sim(x, t)}{\partial t} = \min\{0, H^*(x, \frac{\partial J^\sim(x, t)}{\partial x})\} \tag{6.20a}$$

$$J^\sim(x, 0) = k(x) \tag{6.20b}$$

Assume that (6.20) has a differentiable solution $J^\sim(x, t)$ which converges for $t \to -\infty$ to a function $\bar{J}$. Then the set

$$N^*(F) = \{x \mid \bar{J}(x) \geq 0\}$$

is the set of initial states of trajectories that, under appropriate control action, will remain in F for all time. Furthermore, the actual construction of the controller that enforces this property needs only to be performed on the *boundary* of $N^*(F)$.

Although in many cases the (numerical) solution of the Hamilton-Jacobi equations (6.18) or (6.20) is a formidable task, it provides a systematic approach to checking safety properties for continuous-time systems, and a starting point for checking safety properties for certain classes of hybrid systems, see e.g. [99] for two worked examples.

6.2   Switching control

In some sense, the use of hybrid controllers for continuous-time systems is classical. Indeed, we can look at *variable structure control, sliding mode control, relay control, gain scheduling* and even *fuzzy control* as examples of hybrid control schemes. The common characteristic of all these control schemes is their *switching* nature; on the basis of the evolution of the plant (the continuous-time system to-be-controlled) and/or the progress of time the hybrid controller switches from one control regime to another.

6.2.1   Switching logic

One way of formalizing *switching control* for a continuous-time input-state-output system is by means of Figure 6.1. Here the supervisor has to decide on the basis of the input and output signals of the system, and possibly external (reference) signals or the progress of time, *which* of the, ordinary, continuous-time controllers is applied to the system. The figure indicates a finite number of controllers; however, we could also have a countable set of controllers. Of course there could in fact be a continuum of controllers, but then the resulting control schemes would no longer be referred to as switching control schemes.

In many cases the different controllers can all be given the same state space (shared state variables), which leads to the simpler switching control structure given in Figure 6.2. In this case the supervisor generates a discrete symbol σ corresponding to a controller $\Sigma_c(\sigma)$ producing the control signal u_σ. An example of a switching control scheme was already encountered in Subsection 2.2.8 (the supervisor model), in which case the supervisor is a finite automaton which produces control signals on the basis of the measurement of the output of the plant and the discrete state of the automaton.

The supervisor in Figures 6.1 and 6.2 is sometimes called a *switching logic*. One of the main problems in the design of a switching logic is that usually it is not desirable to have *"chattering"*, that is, very fast switching. There are basically two ways to suppress chattering: one is sometimes called *hysteresis switching logic* and the other *dwell-time switching logic*.

- (Hysteresis switching logic).

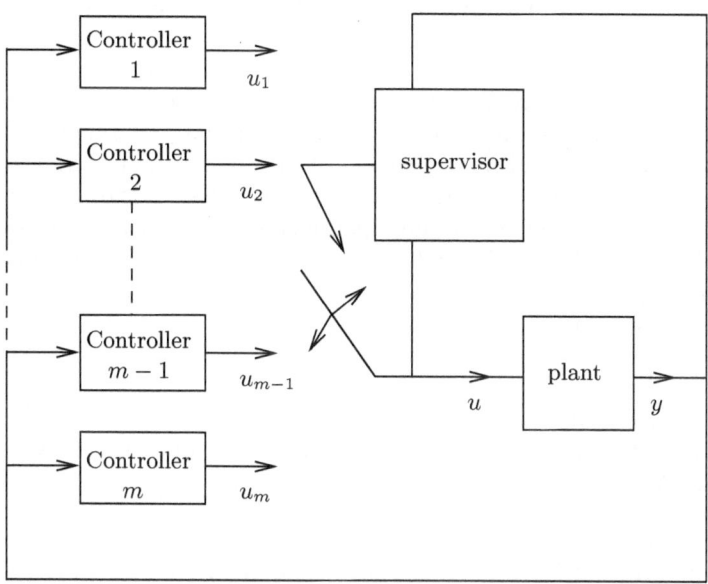

Figure 6.1: Switching control

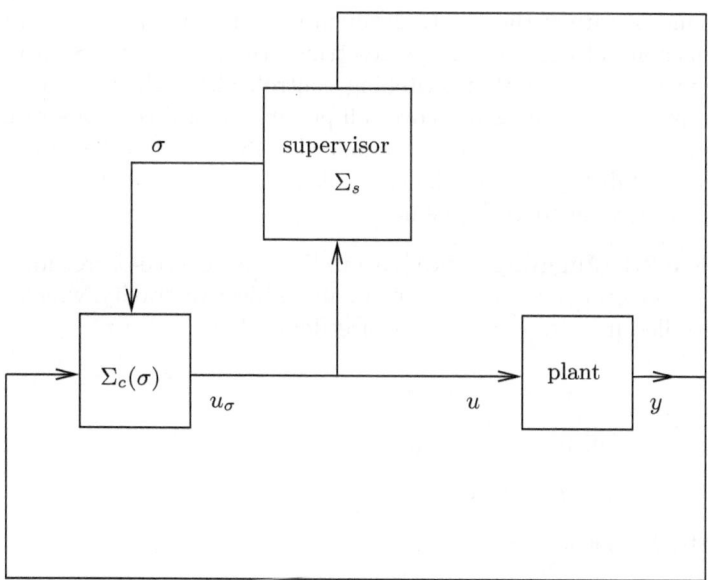

Figure 6.2: Switching control with shared controller state

Choose $h > 0$. Suppose at some time t_0, the supervisor Σ_s has just changed the value of σ to q. Then σ is held *fixed* at the value q unless and until there exists some $t_1 > t_0$ at which $\pi_p + h < \pi_q$ (or $(1 + h)\pi_p < \pi_q$, if a scale-invariant criterion is needed) for some p. If this occurs then σ is set equal to p. Here π_σ denotes some performance criterium depending on σ and some signal derived from the input and the output signals of the continuous-time plant, which is used in the switching logic. Clearly, because of the treshold parameter h no infinitely fast switching will occur. (The idea is similar to the use of a *boundary layer* around the switching surface in order to avoid chattering in sliding mode control; see Subsection 6.2.3.)

- (Dwell-time switching logic).

  The basic idea is to have some fixed time $\tau > 0$ (called the dwell-time) such that, once a symbol σ is chosen by the supervisor it will remain constant for at least a time equal to τ. There are many versions of dwell-time switching logics.

We refer to [115] for further developments and other switching logic schemes. From a general hybrid system point of view a continuous-time system (plant) controlled by switching logic is a hybrid system with continuous state corresponding to the continuous state of the plant and the continuous state of the controller, together with discrete states residing in the controller. Usually, the transition from one discrete state to another will *not* give rise to a jump in the continuous state of the plant, although it could give rise to a jump in the continuous state of the controller ("resetting" the controller). Seen from this point of view it is clear that switching control, although being a very general concept, certainly does not cover all possible hybrid control strategies for continuous-time systems. In fact, the following hybrid control example (taken from [163]) exhibits jumps in the controlled plant and therefore does not fit within the switching control scheme.

Example 6.2.1 (Juggling robot). Consider a one degree-of-freedom juggler, i.e., a system composed of an object (a ball) subject to gravity, which bounces on a controlled mass (a one degree-of-freedom robot).

$$
\begin{aligned}
m_1 \ddot{y}_1 &= -m_1 g \\
m_2 \ddot{y}_2 &= -m_2 g + u \\
y_1 - y_2 &\geq 0 \\
\dot{y}_1(t_k^+) - \dot{y}_2(t_k^+) &= -e[\dot{y}_1(t_k^-) - \dot{y}_2(t_k^-)]
\end{aligned}
\tag{6.21}
$$

where $e \in [0, 1]$ is Newton's restitution coefficient (see also Subsection 2.2.3). Here y_1 represents the height of the ball, while y_2 represents the vertical coordinate of the juggling robot (which is controlled by u). The only way of influencing the trajectory of the ball is through the *impacts*, i.e., at the times

t_k such that $y_1(t_k) - y_2(t_k) = 0$, when the velocity of the ball is reset depending on its initial velocity and the initial velocity of the robot. Note that in order to compute the velocity of the ball (and of the robot) just after an impact time one also needs to invoke the law of conservation of momentum, that is,

$$m_1 \dot{y}_1(t_k^-) + m_2 \dot{y}_2(t_k^-) = m_1 \dot{y}_1(t_k^+) + m_2 \dot{y}_2(t_k^+).$$

6.2.2 PWM control

In some applications one encounters control systems of the following form:

$$\dot{x} = f(x, u), \quad x \in X, \quad u \in U, \tag{6.22}$$

where X is some continuous space, say $\mathbb{R}^n$, while the input space U is a *finite* space (or, more generally, the product of a continuous and a finite space). An appealing class of examples consists of *power converters*, see e.g. Subsection 2.2.13, where the discrete inputs correspond to the switches being closed or open. Since the input space is finite such systems can be considered as a special case of hybrid systems.

In some cases, for instance in power converters, it makes sense to relate the behavior of these hybrid systems to the behavior of an associated control system with *continuous* inputs, using the notion of *Pulse Width Modulation*. Consider for concreteness a control system with $U = \{0, 1\}$, that is,

$$\dot{x} = f(x, u), \quad u \in \{0, 1\}. \tag{6.23}$$

The Pulse Width Modulation (PWM) control scheme for (6.23) can be described as follows. Consider a so-called *duty cycle* of fixed small duration Δ. On every duty cycle the input u is switched exactly one time from 1 to 0. The fraction of the duty cycle on which the input holds the fixed value 1 is known as the *duty ratio* and is denoted by α. The duty ratio may also depend on the state x (or, more precisely, on the value of the state sampled at the beginning of the duty cycle). On every duty cycle $[t, t + \Delta]$ the input is therefore defined by

$$
\begin{aligned}
u(\tau) &= 1, \quad \text{for } t \leq \tau < t + \alpha\Delta \\
u(\tau) &= 0, \quad \text{for } t + \alpha\Delta \leq \tau < t + \Delta
\end{aligned}
\tag{6.24}
$$

It follows that the state x at the end of this duty cycle is given by

$$x(t + \Delta) = x(t) + \int_t^{t+\alpha\Delta} f(x(\tau), 1)d\tau + \int_{t+\alpha\Delta}^{t+\Delta} f(x(\tau), 0)d\tau \tag{6.25}$$

The ideal averaged model of the PWM controlled system is obtained by letting the duty cycle duration Δ go to zero. In the limit, the above formula (6.25) then yields

$$\dot{x}(t) = \lim_{\Delta \to 0} \frac{x(t + \Delta) - x(t)}{\Delta} = \alpha f(x(t), 1) + (1 - \alpha)f(x(t), 0) \tag{6.26}$$

The duty ratio α can be thought of as a *continuous-valued input*, taking its values in the *interval* $[0,1]$. Hence for sufficiently small Δ the trajectories of the continuous-time system (6.26) will be close to trajectories of the hybrid system (6.23). Note that in general the full behavior of the hybrid system (6.23) will be richer than that of (6.26); not every trajectory of (6.23) can be approximated by trajectories of (6.26).

The Pulse Width Modulation control scheme originates from the control of switching power converters, where usually it is reasonable to assume that the switches can be opened ($u = 0$) and closed ($u = 1$) sufficiently fast at any ratio $\alpha \in [0,1]$.

A similar analysis can be performed for a control system (6.23) with *arbitrary* finite input space $U \subset \mathbb{R}^m$. Then the PWM-associated continuous-time system has continuous inputs α taking values in the *convex hull* of U.

6.2.3   Sliding mode control

A classical example of switching control is *variable structure control*, as already partly discussed in Subsection 2.2.7 and Chapter 3. Consider a control system described by equations of the form

$$\dot{x}(t) = f(x(t), u(t))$$

where u is the (scalar) control input. Suppose that a switching control scheme is employed that uses a state feedback law $u(t) = \phi_+(x(t))$ when the scalar variable $y(t)$ defined by $y(t) = h(x(t))$ is positive and a feedback $u(t) = \phi_-(x(t))$ whenever $y(t)$ is negative. Writing $f_+(x) = f(x, \phi_+(x))$ and $f_-(x) = f(x, \phi_-(x))$, we obtain a dynamical system that obeys the equation $\dot{x}(t) = f_+(x(t))$ on the subset where $h(x)$ is positive, and that follows $\dot{x}(t) = f_-(x(t))$ on the subset where $h(x)$ is negative. The surface $\{x \mid h(x) = 0\}$ is called the *switching surface*. The equations can be taken together in the form (see (3.3))

$$\dot{x} = \tfrac{1}{2}(1 + v)f_+(x) + \tfrac{1}{2}(1 - v)f_-(x), \quad v = \operatorname{sgn}(h(x)) \tag{6.27}$$

The extension to multivariable inputs u and outputs y is straightforward. Solutions in Filippov's sense (cf. Chapter 3) can be used in the situation in which there is a "chattering" behavior around the switching surface $h(x) = 0$. The resulting solutions will remain on the switching surface for some time. For systems of the form (6.27), where v is the input, the different solution notions discussed by Filippov all lead to the same solution, which is most easily described using the concept of the *equivalent control* $v_{\text{eq}}(x)$. The equivalent control is defined as the function of the state x, with values in the interval $[-1, 1]$, that is such that the vector field defined by

$$\dot{x}(t) = \tfrac{1}{2}(1 + v_{\text{eq}}(x))f_+(x) + \tfrac{1}{2}(1 - v_{\text{eq}}(x))f_-(x)$$

leaves the switching surface $h(x) = 0$ invariant (see Chapter 3 for further discussion).

In *sliding mode control*, or briefly, *sliding control*, the theory of variable structure systems is actually taken as the starting point for control design. Indeed, some scalar variable s depending on the state x and possibly the time t is considered such that the system has a "desired behavior" whenever the constraint $s = 0$ is satisfied. The set $s = 0$ is called the *sliding surface*. Now a control law u is sought such that the following *sliding condition* is satisfied (see e.g. [144])

$$\frac{1}{2}\frac{d}{dt}s^2 \leq -\alpha|s| \tag{6.28}$$

where α is a strictly positive constant. Essentially, (6.28) states that the squared "distance" to the sliding surface, as measured by s^2, decreases along all system trajectories. Thus it constraints trajectories to point towards the sliding surface. Note that the lefthand side can be also written as $s\dot{s}$; a more general sliding condition can therefore be formulated as designing u such that $s\dot{s} < 0$ for $s \neq 0$.

The methodology of sliding control can be best demonstrated by giving a typical example.

Example 6.2.2. Consider the second-order system

$$\ddot{q} = f(q, \dot{q}) + u$$

where u is the control input, q is the scalar variable of interest (e.g. the position of a mechanical system), and the dynamics described by f (possibly non-linear or time-varying) is not exactly known, but estimated by $\hat{f}$. The estimation error on f is assumed to be bounded by some known function $F(q, \dot{q})$:

$$|\hat{f} - f| \leq F.$$

In order to have the system track some desired trajectory $q(t) = q_d(t)$, we define a sliding surface $s = 0$ with

$$s = \dot{e} + \lambda e$$

where $e = q - q_d$ is the tracking error, and the constant $\lambda > 0$ is such that the error dynamics $s = 0$ has desirable properties (e.g. sufficiently fast convergence of the error to zero). We then have

$$\dot{s} = f + u - \ddot{q}_d + \lambda\dot{e}.$$

The best approximation $\hat{u}$ of a continuous control law that would achieve $\dot{s} = 0$ is therefore

$$\hat{u} = -\hat{f} + \ddot{q}_d - \lambda\dot{e}.$$

Note that $\hat{u}$ can be interpreted as our best estimate of the equivalent control. In order to satisfy the sliding condition (6.28) we add to $\hat{u}$ a term that is *discontinuous* across the sliding surface:

$$u = \hat{u} - k\,\text{sgn}\,s \tag{6.29}$$

where sgn is the signum function defined in (1.13). By choosing k in (6.29) large enough we can guarantee that the sliding condition (6.28) is satisfied. Indeed, we may take $k = F + \alpha$. Note that the control discontinuity k across the sliding surface $s = 0$ increases with the extent of parametric uncertainty F.

The occurrence of a sliding mode may not be desirable from an engineering point of view; depending on the actual implementation of the switching mechanism, a quick succession of switches may occur which may lead to increased wear and to high-frequency vibrations in the system. Hence for the actual implementation of sliding mode control it is usually needed, in order to avoid very fast switching, to embed the sliding surface in a thin *boundary layer*, such that switching will only occur outside this boundary layer (hysteresis switching logic). Furthermore, the discontinuity sgn s in the control law can be further smoothened, say, by replacing sgn by a steep sigmoid function. Of course, such modifications may deteriorate the performance of the closed-loop system.

One of the main advantages of sliding mode control, in addition to its conceptual simplicity, is its robustness with respect to uncertainty in the system data. A possible disadvantage is the excitation of unmodeled high-frequency modes.

6.2.4 Quadratic stabilization by switching control

In continuation of our discussion of quadratic stability in Chapter 5, let us now consider the problem of *quadratic stabilization* of a multi-modal linear system

$$\dot{x} = A_i x, \quad i \in I, \quad x \in \mathbb{R}^n \tag{6.30}$$

where I is some finite index set. The problem is to find a switching rule such that the controlled system has a single quadratic Lyapunov function $x^T P x$. It is not difficult to show that this problem can be solved if there exists a convex combination of the matrices A_i, $i \in I$, which is *Hurwitz* (compare with the corresponding discussion in Chapter 5, Subsection 5.2.3). To verify this claim, assume that the matrix

$$A := \sum \alpha_i A_i \quad (\alpha_i \geq 0, \ \textstyle\sum \alpha_i = 1) \tag{6.31}$$

is Hurwitz. Take a positive definite matrix Q, and let P be the solution of the Lyapunov equation $A^T P + PA = -Q$; because A is Hurwitz, P is positive definite. Let now x be an arbitrary nonzero vector. From $x^T(A^T P + PA)x < 0$ it follows that

$$\sum_i \alpha_i [x^T (A_i^T P + PA_i)x] \ < \ 0. \tag{6.32}$$

Because all the α_i are nonnegative, it follows that at least one of the numbers $x^T(A_i^T P + PA_i)x$ must be negative, and in fact we obtain the stronger statement

$$\bigcup_{i \in I} \{x \mid x^T(A_i^T P + PA_i)x \leq -\frac{1}{N}x^T Q x\} = \mathbb{R}^n \tag{6.33}$$

where N denotes the number of modes.

It is clear how a switching rule for (6.30) may be chosen such that asymptotic stability is achieved; for instance the rule

$$i(x) := \arg\min x^T(A_i^T P + PA_i)x \tag{6.34}$$

is an obvious choice.

The minimum rule (6.34) indeed leads to an asymptotically stable system, which is well-defined if one extends the number of discrete states so as to include possible sliding modes. To avoid sliding modes, the minimum rule may be adapted so that the regions in which different modes are active will overlap rather than just touch. For instance, a modified switching rule (based on hysteresis switching logic) may be chosen which is such that the system will stay in mode i as long as the continuous state x satisfies

$$x^T(A_i^T P + PA_i)x \leq -\frac{1}{2N}x^T Q x. \tag{6.35}$$

When the bound in (6.35) is reached, a switch will take place to a new mode j that may for instance be determined by the minimum rule. At the time at which the new mode j is entered, the number $x^T(A_j^T P + PA_j)x$ must be less than or equal to $-\frac{1}{N}x^T Q x$. Suppose that the switch to mode j occurs at continuous state x_0. The time until the next mode switch is given by

$$\tau_j(x_0) = \min\{t \geq 0 \mid x_0^T e^{tA_j^T}[A_j^T P + PA_j + \frac{1}{2N}Q]e^{tA_j}x_0 \geq 0\} \tag{6.36}$$

(taken to be infinity if the set over which the minimum is taken is empty). Note that $\tau_j(x_0)$ is positive, its dependence on x_0 is lower semi-continuous, and satisfies $\tau_j(\alpha x_0) = \tau_j(x_0)$ for all nonzero real α. Therefore there is a positive lower bound to the time the system will stay in mode j, namely

$$T_j := \min\{\tau_j(x_0) \mid \|x_0\| = 1, \ x_0^T(A_i^T P + PA_i)x_0 = -\frac{1}{2N}x_0^T Q x_0\}. \tag{6.37}$$

So under this switching strategy the system is asymptotically stable and there will be no chattering.

6.3 Hybrid feedback stabilization

In this section we treat a few examples of (simple) continuous-time systems that can be stabilized by hybrid feedback, so as to illustrate the potential of hybrid feedback schemes.

6.3.1 Energy decrease by hybrid feedback

First we consider the well-known *harmonic oscillator*, with equations given as (after a normalization of the constants)

$$
\begin{aligned}
\dot{q} &= v \\
\dot{v} &= -q + u \\
y &= q
\end{aligned}
\tag{6.38}
$$

It is well-known that the harmonic oscillator cannot be asymptotically stabilized using static output feedback. However, since the system is controllable and observable, it can be asymptotically stabilized using *dynamic* output feedback. (One option is to build an observer for the system, and to substitute the estimate of the state produced by this observer into the asymptotically stabilizing *state* feedback law.) As an illustration of the novel aspects created by *hybrid* control strategies, we will show (based on [9]) that the harmonic oscillator can be also asymptotically stabilized by a *hybrid static output* feedback.

Example 6.3.1. Consider the following controller automaton for the harmonic oscillator. It consists of two locations, denoted $+$ and $-$. In each location we define a continuous static output feedback $u_+(q)$, respectively $u_-(q)$ by

$$
u_+(q) = \begin{cases} 0 & \text{for } q \geq 0 \\ -3q & \text{for } q < 0 \end{cases}
\tag{6.39}
$$

and

$$
u_-(q) = \begin{cases} -3q & \text{for } q \geq 0 \\ 0 & \text{for } q < 0 \end{cases}
\tag{6.40}
$$

Furthermore, we define a *dwell-time* T_+, respectively T_- for each location, which specifies how long the controller will remain in a location before a transition is made, either to the other or to the same location. The value of q at the transition time is denoted by q_{tr}. The controller automaton is depicted in Figure 6.3.

To verify that such a hybrid feedback asymptotically stabilizes (6.38), consider the energy function $H(q, p) = \frac{1}{2}(q^2 + v^2)$. In the region where $u = 0$ there is no change in the energy. When $u(q) = -3q$, the time-derivative of the energy along the solution is

$$
\frac{d}{dt}(\frac{1}{2}(q^2(t) + v^2(t))) = -3q(t)v(t)
$$

so that the energy decreases if $q \cdot v > 0$. As shown in Figure 6.3, after a transition the input $u = -3q$ is in operation only when the sign of $q(t) \cdot v(t)$

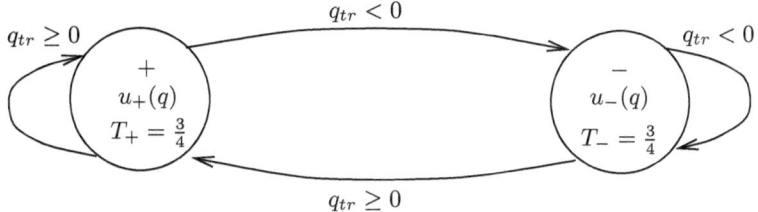

Figure 6.3: Hybrid feedback for the harmonic oscillator

changes from negative to positive, and then for at most $\frac{3}{4}$ time units, in which $q(t) \cdot v(t)$ remains positive. (Note that with $u = -3q$ the solutions $q(t)$ and $v(t)$ of the closed-loop system are shifted sinusoids with frequency 2; hence the product $q(t) \cdot v(t)$ is a shifted sinusoid with frequency 4. However, since $\frac{\pi}{4} > \frac{3}{4}$, $q(t) \cdot v(t)$ will not change sign during the time the input $u = -3q$ is in operation.)

An alternative asymptotically stabilizing hybrid feedback, this time having *three* locations is provided by the following controller automaton depicted in Figure 6.4, with $\delta > 0$ small. In order to verify that this hybrid feed-

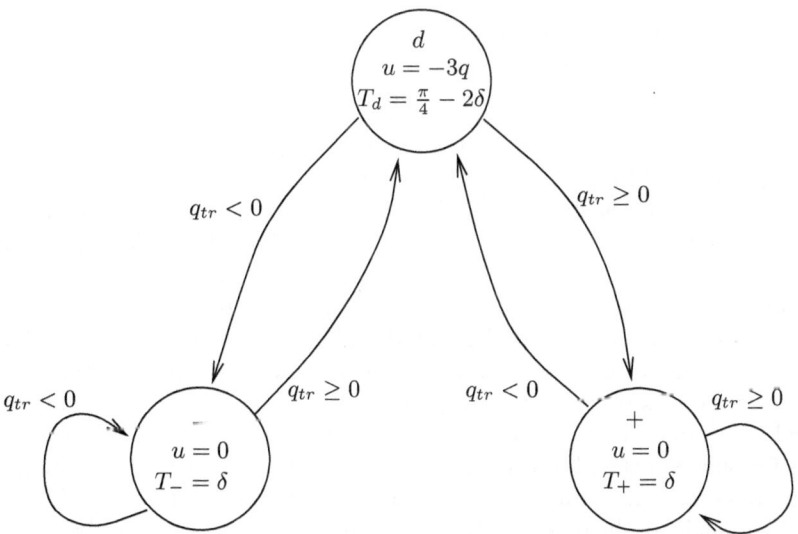

Figure 6.4: An alternative hybrid feedback

back also stabilizes the harmonic oscillator, consider again the total energy $H(q,v) = \frac{1}{2}(q^2 + v^2)$. At the locations $+$ and $-$ there is no change in energy. These two locations serve to detect a shift of q from positive to negative in the case of location $+$, and from negative to positive in the case of location

−. In both cases the change in sign will occur within a bounded number of transition periods. Then the location is shifted to d, which serves as a *dissipation* location. Furthermore, since during both + and − the trajectory turns clockwise, when d is switched on, the initial condition starts at either the first or the third orthant (i.e. $q \cdot v > 0$), and since $T_d = \frac{\pi}{4} - 2\delta$, the solution stays in that orthant as long as d is on. Furthermore, when the location d is reached after a transition, there is a loss of order of $\frac{3}{4}$ of the energy (up to δ) before one of the locations + or − is switched on, depending on whether $q \geq 0$ or $q < 0$, and the search for the change of the sign starts again. All in all, once + is on with $q \geq 0$, or − is on with $q < 0$, the hybrid trajectory passes through d infinitely often, with bounded gaps between occurrences, and each time there is a proportional loss of energy of order close to $\frac{3}{4}$. The conclusion is that $H(q,v)$ tends to zero, uniformly for $q(0), v(0)$ in a compact set. This proves the asymptotic stabilization, provided the hybrid initial condition satisfies $q \geq 0$ in location +, or $q < 0$ in location −. It is straightforward to check that starting from an arbitrary initial condition, the hybrid trajectory reaches one of these situations within an interval of length $\frac{\pi}{2}$, and thus asymptotic stability is shown.

The main difference between both hybrid feedback stabilization schemes is that in the first case we had only two locations with a *nonlinear* control, while in the second case we only employ linear feedback at the price of having three locations. Furthermore, one could compare the convergence and robustness properties of both schemes, as well as compare the two schemes with a *dynamic* asymptotically stabilizing output feedback.

As a preliminary conclusion of this subsection we note that the above hybrid feedback stabilization strategies for the harmonic oscillator are based on *energy* considerations. Indeed, the hybrid feedback schemes were constructed in such a manner that the energy is always *non-increasing* along the hybrid trajectories. The difficulty of the actual *proof* of asymptotic stability lies in showing that the energy is indeed in all cases non-increasing, and that moreover eventually all energy is dissipated. (The idea is somewhat similar to LaSalle's version of Lyapunov stability theory: the Lyapunov function should be always non-increasing along trajectories, while there should be no nontrivial invariant sets where the Lyapunov function remains constant.) The tools for showing this, however, are rather *ad hoc*. Furthermore, there is not yet a systematic procedure to construct hybrid feedback stabilization strategies. The main problem seems to be that in the hybrid case we do not have an explicit expression of the energy decrease analogous to the time-derivative $\frac{d}{dt}H = \frac{\partial H}{\partial x}(x)f(x,u)$ of the energy H for a continuous-time system $\dot{x} = f(x,u)$.

Clearly, the idea of a non-increasing energy can be extended from physical energy functions H to general Lyapunov functions V (with the ubiquitous difficulty of *finding* feasible Lyapunov functions). The setting seems somewhat similar to the use of *control Lyapunov functions* in the feedback stabilization of *nonlinear* control systems, see e.g. [143].

6.3.2 Stabilization of nonholonomic systems

While the above example of stabilizing a harmonic oscillator by hybrid static output feedback may not seem very convincing, since the goal could be also reached by other means, we now turn to a different class of systems where the usefulness of hybrid feedback stabilization schemes is immediately clear. Indeed, let us consider the example of the *nonholonomic integrator*

$$
\begin{aligned}
\dot{x} &= u \\
\dot{y} &= v \\
\dot{z} &= xv - yu
\end{aligned}
\tag{6.41}
$$

where u and v denote the controls. To indicate the physical significance of this system it can be shown that any kinematic completely non-holonomic mechanical system with three states and two inputs can be converted into this form (see e.g. [118]).

The nonholonomic integrator is the prototype of a nonlinear system, which is *controllable*, but nevertheless cannot be asymptotically stabilized using *continuous* state feedback (static or dynamic). The reason is that the nonholonomic integrator violates what is called *Brockett's necessary condition* [29], which is formulated in the following theorem.

Theorem 6.3.2. *Consider the control system*

$$
\dot{x} = f(x, u), \quad x \in \mathbb{R}^n, \quad u \in \mathbb{R}^m, \quad f(0, 0) = 0,
\tag{6.42}
$$

where f is a C^1 function. If (6.42) is asymptotically stabilizable (about 0) using a continuous feedback law $u = \alpha(x)$, then the image of every open neighborhood of $(0, 0)$ under f contains an open neighborhood of 0.

Remark 6.3.3. Since we allow for *continuous* feedback laws (instead of C^1 or locally Lipschitz feedback laws) some care should be taken in connection with the *existence and uniqueness* of solutions of the closed-loop system.

It is readily seen that the nonholonomic integrator does not satisfy the condition mentioned in Theorem 6.3.2 (Brockett's necessary condition), despite the fact that it is controllable (as can be rather easily seen). Indeed, $(0, 0, \epsilon) \notin \text{Im}(f)$ for any $\epsilon \neq 0$, and so the nonholonomic integrator cannot be stabilized by a time-invariant continuous feedback.

In fact, the nonholonomic integrator is an example of a whole class of systems, sharing the same property. For example, actuated mechanical systems subject to *nonholonomic kinematic constraints* (like rolling without slipping) do not satisfy Brockett's necessary condition, but are often controllable.

For the nonholonomic integrator we consider the following sliding mode control law (taken from [21]):

$$
\begin{aligned}
u &= -x + y \operatorname{sgn} z \\
v &= -y - x \operatorname{sgn} z
\end{aligned}
\tag{6.43}
$$

Consider a Lyapunov function for the (x, y)-subspace:

$$V(x, y) = \tfrac{1}{2}(x^2 + y^2)$$

The time-derivative of V along the trajectories of the closed-loop system (6.41–6.43) is negative:

$$\dot{V} = -x^2 + xy \operatorname{sgn} z - y^2 - xy \operatorname{sgn} z = -(x^2 + y^2) = -2V. \tag{6.44}$$

It already follows that the variables x, y converge to zero. Now let us consider the variable z. Using equations (6.41–6.43) we obtain

$$\dot{z} = xv - yu = -(x^2 + y^2) \operatorname{sgn} z = -2V \operatorname{sgn} z \tag{6.45}$$

Since V does not depend on z and is a positive function of time, the absolute value of the variable z will thus decrease and will be able to reach zero in finite time provided the inequality

$$2 \int_0^\infty V(\tau)d\tau > |z(0)| \tag{6.46}$$

holds. If this inequality is an equality, then $z(t)$ converges to the origin in *infinite* time. Otherwise, it converges to some constant non-zero value that has the same sign as $z(0)$. After reaching zero $z(t)$ will remain there, since according to (6.45), all trajectories are directed towards the surface $z = 0$ (the sliding surface), while x and y always converge to the origin while within this surface.

From (6.44) it follows that

$$V(t) = V(0)e^{-2t} = \frac{1}{2}(x^2(0) + y^2(0))e^{-2t}.$$

Substituting this expression in (6.46) and integrating, we find that the condition for the system to be asymptotically stable is that

$$\tfrac{1}{2}(x^2(0) + y^2(0)) \geq |z(0)|.$$

The inequality

$$\tfrac{1}{2}(x^2 + y^2) < |z| \tag{6.47}$$

defines a parabolic region in the state space. The above derivation is summarized in the following theorem.

Theorem 6.3.4. *If the initial conditions for the system (6.41) do not belong to the region defined by (6.47), then the control (6.43) asymptotically stabilizes the system.*

If the initial data are inside the parabolic region defined by (6.47), we can use any control law that first steers it outside. In fact, any nonzero constant control can be applied. Namely, if $u = u_0 = $ const, $v = v_0 = $ const, then

$$
\begin{aligned}
x(t) &= u_0 t + x_0 \\
y(t) &= v_0 t + y_0 \\
z(t) &= x(t)v_0 - y(t)u_0 \\
&= t(x_0 v_0 - y_0 u_0) + z_0
\end{aligned}
$$

With such x, y and z the left hand side of (6.47) is quadratic with respect to time t, while the right hand side is linear. Hence, the state will always leave the parabolic region defined by (6.47), and then we switch to the control given by (6.43). Note that the resulting control strategy is inherently hybrid in nature: first we apply the constant control $u = u_0$, $v = v_0$ if the state belongs to the parabolic region defined by (6.47), and if the system is outside this region then we switch to the sliding control (6.43).

A possible drawback of the above hybrid feedback scheme is caused by the application of sliding control: in principle we will get chattering around the sliding surface $z = 0$. Although this can be remedied with the usual tools in sliding control (e.g. putting a boundary layer around the sliding surface, or approximating the sgn function by a steep sigmoid function), it motivates the search for an alternative hybrid feedback scheme that completely avoids the chattering behavior.

Remark 6.3.5. An alternative sliding mode feedback law can be formulated as follows:

$$
\begin{aligned}
u &= -x + \frac{y}{x^2 + y^2}\, \text{sgn}\, z \\
v &= -y - \frac{x}{x^2 + y^2}\, \text{sgn}\, z.
\end{aligned}
\tag{6.48}
$$

This feedback will result in the same equation (6.44) for the time-derivative of V, while the equation (6.45) for the time-derivative of z is improved to

$$
\dot{z} = -\,\text{sgn}\, z. \tag{6.49}
$$

Of course, the price that needs to be paid for this more favorable behavior is that the control (6.48) is unbounded for $x^2 + y^2$ tending to zero.

In connection to the above hybrid feedback schemes (6.43) and (6.48) we consider the following alternative scheme, taken from [77, 78]. Let $q = (x, y, z)$, and define $W(q) = (w_1(q), w_2(q)) \in \mathbb{R}^2$ by

$$
\begin{aligned}
w_1(q) &= z^2 \\
w_2(q) &= x^2 + y^2
\end{aligned}
$$

We now define four regions in $\mathbb{R}^2$, and construct the following hybrid feedback with four locations.

1. Pick four continuous functions $\pi_k : [0, \infty) \to \mathbb{R}$, $k \in I = \{1, 2, 3, 4\}$ with the following properties.

   (a) $\pi_k(0) = 0$ for each $k \in \{1, 2, 3, 4\}$.

   (b) for each $k \in \{1, 2, 3, 4\}$ and for all $w > 0$

   $$0 < \pi_1(w) < \pi_2(w) < \pi_3(w) < \pi_4(w).$$

   (c) π_1 and π_2 are bounded.

   (d) π_1 is such that if $w \to 0$ exponentially fast then $\frac{w}{\pi_1(w)} \to 0$ exponentially fast.

   (e) π_4 is differentiable on some non-empty, half-open interval $(0, c]$ and

   $$\frac{d\pi_4}{dw}(w) < \frac{\pi_4(w)}{w}, \quad w \in (0, c].$$

   Moreover, if $w \to 0$ exponentially fast, then $\pi_4(w) \to 0$ exponentially fast.

2. Partition the closed positive quadrant $\Omega \subset \mathbb{R}^2$ into four overlapping regions

   $$
   \begin{aligned}
   R_1 &= \{(w_1, w_2) \in \Omega : 0 \le w_2 < \pi_2(w_1)\} \\
   R_2 &= \{(w_1, w_2) \in \Omega : \pi_1(w_1) < w_2 < \pi_4(w_1)\} \\
   R_3 &= \{(w_1, w_2) \in \Omega : w_2 > \pi_3(w_1)\} \\
   R_4 &= \{(0, 0)\}.
   \end{aligned}
   $$

3. Define the control law

   $$(u(t), v(t))^T = g_{\sigma(t)}(q(t))$$

   where $\sigma(t)$ is a piecewise constant switching signal taking values in $I = \{1, 2, 3, 4\}$ and

   $$g_1(q) = \begin{bmatrix} 1 \\ 1 \end{bmatrix}$$

   $$g_2(q) = \begin{bmatrix} x + \dfrac{yz}{x^2 + y^2} \\ y - \dfrac{xz}{x^2 + y^2} \end{bmatrix}$$

$$g_3(q) = \begin{bmatrix} -x + \dfrac{yz}{x^2 + y^2} \\ -y - \dfrac{xz}{x^2 + y^2} \end{bmatrix}$$

$$g_4(q) = \begin{bmatrix} 0 \\ 0 \end{bmatrix}.$$

The signal $\sigma(t)$ is defined to be continuous from the right and is determined recursively by

$$\sigma(t) = \phi(q(t), \sigma^-(t)), \quad \sigma^-(t_0) = \sigma_0 \in I$$

where $\sigma^-(t)$ denotes the limit of $\sigma(t)$ from below and $\phi : \mathbb{R}^3 \times I \to I$ is the transition function

$$\phi(q, \sigma) = \begin{cases} \sigma & \text{if } W(q) \in R_\sigma \\ \max\{k \in I : W(q) \in R_k\} & \text{otherwise} \end{cases}$$

A typical example is obtained by taking $\pi_1(w_1) = (1 - e^{-\sqrt{w_1}})$, $\pi_2 = 2\pi_1$, $\pi_3 = 3\pi_1$, and $\pi_4 = 4\pi_1$. The resulting regions R_1, R_2 and R_3 are shown in Figure 6.5.

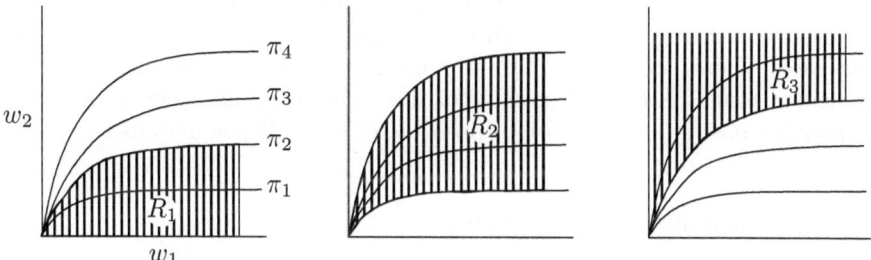

Figure 6.5: A typical partition of Ω

Although the closed-loop system is not globally Lipschitz, global existence of solutions can be easily established. Indeed, simple calculations show that

$$\dot{w}_1 \leq 2w_1 + w_2, \quad \dot{w}_2 \leq 2w_2 + 2.$$

Since the bounds for the right-hand sides of the above equations are globally Lipschitz with respect to w_1 and w_2, these variables and their derivatives are bounded on any finite interval, from which the global existence of solutions can be established (see [78] for details). Note that if we start in region R_1, then we apply the constant control $u = 1$ and $v = 1$ as in the previous hybrid sliding mode design. For the same reason as given above, this will lead us to the region R_2. Furthermore, if the switching signal is equal to 2 or 3 we obtain (compare with (6.49))

$$\dot{w}_1 = -2w_1$$

and so w_1, and hence z, will converge to zero. Detailed analysis (cf. [78]) indeed shows this to happen, and moreover shows that also x and y converge to zero. The convergence of all variables is exponential.

Let us try to summarize what we have learned from this example of hybrid feedback stabilization of a nonholonomic system. As a general remark, there seems to be at the moment no general design methodology for constructing hybrid feedback stabilization schemes. The first scheme for asymptotic stabilization of the nonholonomic integrator that we discussed is based on *sliding control*, while also the second scheme takes much inspiration from such sliding control schemes. However it is not easy to extract a systematic design methodology from this example. Furthermore, the actual proofs that the proposed hybrid feedback schemes do work are typically complicated and rather *ad hoc*.

6.3.3   Set-point regulation of mechanical systems by energy injection

In analogy with Subsection 6.3.1 let us continue with hybrid feedback strategies that are based on *energy* considerations. Instead of *decreasing* the energy of the system in bringing the system to a desired rest configuration, there are also cases where actually we like to "inject" energy into the system in an efficient manner. A typical example of this is to *swing up* a pendulum from its hanging position to the upright position, cf. [10]. The equations of motion of a pendulum are given by

$$J_p\ddot{\theta} - mgl\sin\theta + mul\cos\theta = 0. \tag{6.50}$$

Here m denotes the mass of the pendulum, concentrated at the end, l denotes the length, and J_p is the moment of inertia with respect to the pivot point. The angle between the vertical and the pendulum is denoted by θ, where θ is positive in the clockwise direction. The acceleration due to gravity is g and the horizontal acceleration of the pivot point is u. Note that the linearized systems are controllable, except when $\theta = \frac{\pi}{2}$ or $-\frac{\pi}{2}$, i.e. except when the pendulum is horizontal. One way to swing the pendulum to its upright position is to inject into the system an amount of energy such that the total energy corresponds to the potential energy of the upright position. Once the pendulum is sufficiently close to the upright position one then *switches* to a (say, linear) controller in order to keep the system near this upright equilibrium—so this is already a switching control strategy! The energy of the system is given by

$$E = \frac{1}{2}J_p\dot{\theta}^2 + mgl\cos\theta \tag{6.51}$$

Computing the derivative of E with respect to time we find

$$\frac{dE}{dt} = J_p\dot{\theta}\ddot{\theta} - mgl\dot{\theta}\sin\theta = mul\dot{\theta}\cos\theta \qquad (6.52)$$

It follows from (6.52) that it is easy to control the energy. Only for $\dot{\theta}\cos\theta = 0$, that is, for $\dot{\theta} = 0$, $\theta = \frac{\pi}{2}$, or $\theta = -\frac{\pi}{2}$, controllability of the energy increase is lost. Physically these singular states correspond to the case when the pendulum reverses its velocity or when it is horizontal. Control action is most effective when the angle θ is close to 0 or π and the velocity is large. To increase energy the horizontal acceleration of the pivot u should be positive when the quantity $\dot{\theta}\cos\theta$ is negative. To change the energy as quickly as possible the magnitude of the control signal should be as large as possible.

Let the desired energy be E_0. The following sliding control law is a simple strategy for achieving the desired energy:

$$u = \mathrm{sat}_c(k(E - E_0))\,\mathrm{sgn}(\dot{\theta}\cos\theta) \qquad (6.53)$$

where k is a design parameter. In this expression the function sat_c saturates the control action at the level c. The strategy is essentially a bang-bang strategy for large errors and a proportional control for small errors. For large values of k the strategy (6.53) is arbitrarily close to the strategy that gives the maximum increase of energy.

Let us now discuss strategies for bringing the pendulum from rest in downward position to a stationary position in the upright equilibrium. The potential energy of the pendulum is normalized to be zero at the upright position and is thus equal to $-2mgl$ at the downward position. One way to swing up the pendulum is therefore to control it in such a way that its energy increases from $-2mgl$ to 0. A very simple strategy to achieve this is as follows. Accelerate the pendulum with maximum acceleration in an arbitrary direction and reverse the acceleration when the velocity becomes zero. It can be seen that this strategy is optimal as long as the pendulum does not reach the horizontal position, because it follows from (6.52) that the acceleration should be reversed when the pendulum reaches the horizontal position.

The behavior of the swing depends critically on the maximum acceleration c of the pivot. If c is large enough the pendulum can be brought up in one swing, but otherwise multiple swings are required.

In fact it can be seen that if $c \geq 2g$ then one swing is sufficient by first using the maximum acceleration until the desired energy is obtained, and then setting the acceleration to zero. (Note that with $c = 2g$ this comes down to maximum acceleration until the pendulum is horizontal, and then switching off the acceleration.) However, using more than two switches as above, it can be seen that it is possible to bring the pendulum in upright position if $\frac{4}{3}g \leq c < 2g$. In fact, if $c = \frac{4}{3}g$ then we maximally accelerate till the pendulum is horizontal, and then we maximally *reverse* the acceleration until the desired energy is reached, at which moment we set the acceleration equal to zero.

It is interesting to compare these energy-based switching strategies with *minimum time* strategies. Indeed, from Pontryagin's Maximum principle it follows that minimum time strategies for $|u|$ bounded by c are of bang-bang type as well. It can be shown that these strategies also have a nice interpretation in terms of energy. They will inject energy into the pendulum at a maximum rate and then *remove* energy at maximum rate in such a way that the energy corresponds to the equilibrium energy at the moment when the upright position is reached. In fact, for small values of c the minimum-time strategies produce control signals that at first are identical to those produced by the energy-based strategies. The final part of the control signals, however, is different, because the energy-based control strategies will set the acceleration equal to zero when the desired energy has been obtained, while in the minimum-time strategies there is an "overshoot" in the energy.

In principle, the same reasoning can be used for set-point regulation of other mechanical systems.

6.4   Notes and References for Chapter 6

Section 6.1 is mainly based on [99] and [102]. The theory of controlled invariance for linear systems can be found in [160], [15], and for the nonlinear case e.g. in [121]. In a more general context the property of controlled invariance has been studied as "viability theory", cf. [11]. Conceptually very much related work in the area of discrete-event systems is the work of Wonham and co-workers, see e.g. [132], [149]. For optimal control of hybrid systems, in particular the extension of the Maximum Principle to the hybrid case, we refer to [147]. Subsection 6.2.1 is largely based on [115], while part of Subsection 6.2.3 is based on the exposition of sliding mode control in [144]. There is a substantial body of literature on the topic of quadratic stabilization that we touched upon in Subsection 6.2.4. A few references are [103], [128], and [154] where the switching rule according to (6.35) is suggested. Subsection 6.3.1 is taken from [9]. The first part of Subsection 6.3.2 is based on [21], see also [22]. The second part of Subsection 6.3.2 is based on [78], see also [77]. Finally, Subsection 6.3.3 is based on [10]. Of course, there is much additional work on the control of hybrid systems that we did not discuss here. For further information the reader may refer to the papers in the recent special issues [7] and [116]. A survey of hybrid systems in process control is available in [94].

Bibliography

[1] R. Alur, C. Courcoubetis, N. Halbwachs, T. A. Henzinger, P.-H. Ho, X. Nicollin, A. Olivero, J. Sifakis, and S. Yovine. The algorithmic analysis of hybrid systems. *Theoretical Computer Science*, 138:3–34, 1995.

[2] R. Alur and D. L. Dill. A theory of timed automata. *Theoretical Computer Science*, 126:183–235, 1994.

[3] R. Alur, T. A. Henzinger, and E. D. Sontag, eds. *Hybrid Systems III.* Lect. Notes Comp. Sci., vol. 1066, Springer, Berlin, 1996.

[4] A. Andronov and C. Chaikin. *Theory of Oscillations.* Princeton Univ. Press, Princeton, NJ, 1949.

[5] P. Antsaklis, W. Kohn, A. Nerode, and S. Sastry, eds. *Hybrid Systems II.* Lect. Notes Comp. Sci., vol. 736, Springer, Berlin, 1995.

[6] P. Antsaklis, W. Kohn, A. Nerode, and S. Sastry, eds. *Hybrid Systems IV.* Lect. Notes Comp. Sci., vol. 1386, Springer, Berlin, 1997.

[7] P. Antsaklis and A. Nerode, guest eds. Special Issue on Hybrid Control Systems. *IEEE Trans. Automat. Contr.*, 43(4), April 1998.

[8] P. J. Antsaklis, J. A. Stiver, and M. D. Lemmon. Hybrid system modeling and autonomous control systems. In *Hybrid Systems* (R. L. Grossman, A. Nerode, A. P. Ravn, and H. Rischel eds.), Lect. Notes. Comp. Sci., vol. 736, Springer, New York, 1993, pp. 366–392.

[9] Z. Artstein. Examples of stabilization with hybrid feedback. In *Hybrid Systems III* (R. Alur, T. A. Henzinger, and E. D. Sontag eds.), Lect. Notes Comp. Sci., vol. 1066, Springer, New York, 1996, pp. 173–185.

[10] K. J. Åstrom and K. Furuta. Swinging up a pendulum by energy control. In *Proc. 13th IFAC World Congress, Vol. E*, 1996, pp. 37–42.

[11] J.-P. Aubin. *Viability Theory.* Birkhäuser, Boston, 1991.

[12] J.-P. Aubin and A. Cellina. *Differential Inclusions: Set-valued Maps and Viability Theory.* Springer, Berlin, 1984.

[13] F. Baccelli, G. Cohen, G. J. Olsder, and J. P. Quadrat. *Synchronization and Linearity.* Wiley, New York, 1992.

[14] T. Basar and G. J. Olsder. *Dynamic Non-cooperative Game Theory* (2nd ed.). Academic Press, Boston, 1995.

[15] G. Basile and G. Marro. *Controlled and Conditioned Invariants in Linear System Theory.* Prentice Hall, Englewood Cliffs, NJ, 1992.

[16] A. Bemporad and M. Morari. Control of systems integrating logic, dynamics, and constraints. *Automatica,* 35:407–428, 1999.

[17] A. Benveniste. Compositional and uniform modelling of hybrid systems. In *Proc. 35th IEEE Conf. Dec. Contr.,* Kobe, Japan, Dec. 1996, pp. 153–158.

[18] A. Benveniste. Compositional and uniform modelling of hybrid systems. *IEEE Trans. Automat. Contr.,* 43:579–584, 1998.

[19] A. Benveniste and P. Le Guernic. Hybrid dynamical systems theory and the SIGNAL language. *IEEE Trans. Automat. Contr.,* 35:535–546, 1990.

[20] F. Black and M. Scholes. The pricing of options and corporate liabilities. *Journal of Political Economy,* 81:637–659, 1973.

[21] A. Bloch and S. Drakunov. Stabilization of a nonholonomic system via sliding modes. In *Proc. 33rd IEEE Conf. Dec. Contr.,* Lake Buena Vista, Dec. 1994, pp. 2961–2963.

[22] A. Bloch and S. Drakunov. Stabilization and tracking in the nonholonomic integrator via sliding modes. *Syst. Contr. Lett.,* 29:91–99, 1996.

[23] W. M. G. van Bokhoven. *Piecewise Linear Modelling and Analysis.* Kluwer, Deventer, the Netherlands, 1981.

[24] D. Bosscher, I. Polak, and F. Vaandrager. Verification of an audio control protocol. In *Formal Techniques in Real-Time and Fault-Tolerant Systems* (H. Langmaack, W.-P. de Roever, and J. Vytopil eds.), Lect. Notes Comp. Sci., vol. 863, Springer, Berlin, 1994.

[25] S. P. Boyd, L. El Ghaoui, E. Feron, V. Balakrishnan. *Linear Matrix Inequalities in System and Control Theory.* SIAM Studies in Applied Mathematics, vol. 15, SIAM, Philadelphia, 1994.

[26] M. S. Branicky, V. S. Borkar, and S. K. Mitter. A unified framework for hybrid control. In *Proc. 33rd IEEE Conf. Dec. Contr.,* Lake Buena Vista, Dec. 1994, pp. 4228–4234.

[27] F. Breitenecker and I. Husinsky, eds. *Comparison of Simulation Software.* EUROSIM — Simulation News, no. 0–19, 1990–1997.

[28] K. E. Brenan, S. L. Campbell, and L. R. Petzold. *Numerical Solution of Initial-Value Problems in Differential-Algebraic Equations*. North-Holland, Amsterdam, 1989.

[29] R. W. Brockett. Asymptotic stability and feedback stabilization. In *Differential Geometric Control Theory* (R. W. Brockett, R. Millman, and H. J. Sussmann eds.), Birkhäuser, Boston, 1983.

[30] R. W. Brockett. Hybrid models for motion control systems. In *Essays on Control: Perspectives in the Theory and its Applications* (H. L. Trentelman and J. C. Willems eds.), Progress Contr. Syst. Th., vol. 14, Birkhäuser, Boston, pp. 29–53, 1993.

[31] B. Brogliato. *Nonsmooth Impact Mechanics. Models, Dynamics and Control.* Lect. Notes Contr. Inform. Sci., vol. 220, Springer, Berlin, 1996.

[32] C. J. Budd. Non-smooth dynamical systems and the grazing bifurcation. In *Nonlinear Mathematics and its Applications* (P. J. Aston ed.), Cambridge Univ. Press, Cambridge, 1996, pp. 219–235.

[33] M. K. Çamlıbel and J. M. Schumacher. Well-posedness of a class of piecewise linear systems. In *Proc. European Control Conference '99*, Karlsruhe, Aug. 31 – Sept. 3, 1999.

[34] M. K. Çamlıbel, W. P. M. H. Heemels, and J. M. Schumacher. The nature of solutions to linear passive complementarity systems. In *Proc. 38th IEEE Conf. Dec. Contr.*, Phoenix, Arizona, Dec. 1999.

[35] Z. Chaochen, C. A. R. Hoare, and A. P. Ravn. A calculus of durations. *Inform. Process. Lett.*, 40:269–276, 1991.

[36] C. Chase, J. Serrano, and P. J. Ramadge. Periodicity and chaos from switched flow systems: contrasting examples of discretely controlled continuous systems. *IEEE Trans. Automat. Contr.*, 38:70–83, 1993.

[37] H. Chen and A. Mandelbaum. Leontief systems, RBV's and RBM's. In *Applied Stochastic Analysis* (M. H. A. Davis and R. J. Elliott eds.), Stochastics Monographs, vol. 5, Gordon & Breach, New York, 1991, pp. 1–43.

[38] L. O. Chua, M. Komuro, and T. Matsumoto. The double scroll family. *IEEE Trans. Circuits Syst.*, 33:1072–1118, 1986.

[39] R. W. Cottle, J.-S. Pang, and R. E. Stone. *The Linear Complementarity Problem*. Academic Press, Boston, 1992.

[40] M. H. A. Davis. *Markov Models and Optimization*. Chapman and Hall, London, 1993.

[41] C. Daws, A. Olivero, S. Tripakis, and S. Yovine. The tool KRONOS. In *Hybrid Systems III. Verification and Control* (E. D. Sontag, R. Alur, and T. A. Henzinger eds.), Lect. Notes Comp. Sci., vol. 1066, Springer, Berlin, 1996, pp. 208–219.

[42] J. H. B. Deane and D. C. Hamill. Chaotic behavior in current-mode controlled DC-DC convertor. *Electron. Lett.*, 27:1172–1173, 1991.

[43] I. Demongodin and N. T. Koussoulas. Differential Petri nets: representing continuous systems in a discrete-event world. *IEEE Trans. Automat. Contr.*, 43:573–579, 1998.

[44] B. De Schutter and B. De Moor. Minimal realization in the max algebra is an extended linear complementarity problem. *Syst. Contr. Lett.*, 25:103–111, 1995.

[45] B. De Schutter and B. De Moor. The linear dynamic complementarity problem is a special case of the extended linear complementarity problem. *Syst. Contr. Lett.*, 34:63–75, 1998.

[46] C. A. Desoer and E. S. Kuh. *Basic Circuit Theory*. McGraw-Hill, New York, 1969.

[47] M. Di Bernardo, F. Garofalo, L. Glielmo, and F. Vasca. Switchings, bifurcations, and chaos in DC/DC converters. *IEEE Trans. Circuits Syst. I*, 45:133–141, 1998.

[48] P. Dupuis. Large deviations analysis of reflected diffusions and constrained stochastic approximation algorithms in convex sets. *Stochastics*, 21:63–96, 1987.

[49] P. Dupuis and A. Nagurney. Dynamical systems and variational inequalities. *Annals of Operations Research*, 44:9–42, 1993.

[50] H. Elmqvist, F. Boudaud, J. Broenink, D. Brück, T. Ernst, P. Fritzon, A. Jeandel, K. Juslin, M. Klose, S. E. Mattsson, M. Otter, P. Sahlin, H. Tummescheit, and H. Vangheluwe. *Modelica™ — a unified object-oriented language for physical systems modeling*. Version 1, September 1997.
www.Dynasim.se/Modelica/Modelica1.html.

[51] G. Escobar, A. J. van der Schaft, and R. Ortega. A Hamiltonian viewpoint in the modelling of switching power converters. *Automatica*, 35:445–452, 1999.

[52] X. Feng, K. A. Loparo, Y. Ji, and H. J. Chizek. Stochastic stability properties of jump linear systems. *IEEE Trans. Automat. Contr.*, 37:38–53, 1992.

[53] A. F. Filippov. Differential equations with discontinuous right-hand side. *Mat. Sb. (N. S.)*, 51 (93):99–128, 1960. (Russian. Translated in: Math. USSR-Sb.)

[54] A. F. Filippov. *Differential Equations with Discontinuous Righthand Sides*. Kluwer, Dordrecht, 1988.

[55] F. R. Gantmacher. *The Theory of Matrices, Vol. I*. Chelsea, New York, 1959.

[56] F. R. Gantmacher. *The Theory of Matrices, Vol. II*. Chelsea, New York, 1959.

[57] A. H. W. Geerts and J. M. Schumacher. Impulsive-smooth behavior in multimode systems. Part I: State-space and polynomial representations. *Automatica*, 32:747–758, 1996.

[58] R. I. Grossman, A. Nerode, A. P. Ravn, and H. Rischel, eds. *Hybrid Systems*. Lect. Notes Comp. Sci., vol. 736, Springer, Berlin, 1993.

[59] J. Guckenheimer and P. Holmes. *Nonlinear Oscillations, Dynamical Systems, and Bifurcations of Vector Fields*. Springer, New York, 1983.

[60] W. Hahn. *Stability of Motion*. Springer, Berlin, 1967.

[61] E. Hairer and G. Wanner. *Solving Ordinary Differential Equations II: Stiff and Differential-Algebraic Problems*. Springer, Berlin, 1991.

[62] N. Halbwachs, P. Caspi, P. Raymond, and D. Pilaud. The synchronous dataflow programming language LUSTRE. *Proc. IEEE*, 79:1305–1320, 1991.

[63] I. Han and B. J. Gilmore. Multi-body impact motion with friction. Analysis, simulation, and experimental validation. *ASME J. of Mechanical Design*, 115:412–422, 1993.

[64] P. T. Harker and J.-S. Pang. Finite-dimensional variational inequality and nonlinear complementarity problems: A survey of theory, algorithms and applications. *Math. Progr. (Ser. B)*, 48:161–220, 1990.

[65] R. F. Hartl, S. P. Sethi, and R. G. Vickson. A survey of the maximum principles for optimal control problems with state constraints. *SIAM Review*, 37:181–218, 1995.

[66] M. L. J. Hautus. The formal Laplace transform for smooth linear systems. In *Mathematical Systems Theory* (G. Marchesini and S. K. Mitter eds.), Lect. Notes Econ. Math. Syst., vol. 131, Springer, New York, 1976, pp. 29–47.

[67] M. L. J. Hautus and L. M. Silverman. System structure and singular control. *Lin. Alg. Appl.*, 50:369–402, 1983.

[68] W. P. M. H. Heemels. *Linear Complementarity Systems: A Study in Hybrid Dynamics.* Ph. D. dissertation, Eindhoven Univ. of Technol., Nov. 1999.

[69] W. P. M. H. Heemels, J. M. Schumacher, and S. Weiland. *Linear complementarity systems.* Internal Report 97 I/01, Dept. of EE, Eindhoven Univ. of Technol., July 1997. To appear in *SIAM J. Appl. Math.* www.cwi.nl/~jms/PUB/ARCH/lcs_revised.ps.Z (revised version).

[70] W. P. M. H. Heemels, J. M. Schumacher, and S. Weiland. The rational complementarity problem. *Lin. Alg. Appl.*, 294:93–135, 1999.

[71] W. P. M. H. Heemels, J. M. Schumacher, and S. Weiland. *Projected dynamical systems in a complementarity formalism.* Technical Report 99 I/03, Measurement and Control Systems, Dept. of EE, Eindhoven Univ. of Technol., 1999.

[72] W. P. M. H. Heemels, J. M. Schumacher, and S. Weiland. Applications of complementarity systems. In *Proc. European Control Conference '99*, Karlsruhe, Aug. 31 – Sept. 3, 1999.

[73] W. P. M. H. Heemels, J. M. Schumacher, and S. Weiland. Well-posedness of linear complementarity systems. In *Proc. 38th IEEE Conf. Dec. Contr.*, Phoenix, Arizona, Dec. 1999.

[74] M. F. Heertjes, M. J. G. van de Molengraft, J. J. Kok, R. H. B. Fey, and E. L. B. van de Vorst. Vibration control of a nonlinear beam system. In *IUTAM Symposium on Interaction between Dynamics and Control in Advanced Mechanical Systems* (D. H. van Campen ed.), Kluwer, Dordrecht, the Netherlands, 1997, pp. 135–142.

[75] U. Helmke and J. B. Moore. *Optimization and Dynamical Systems.* Springer, London, 1994.

[76] T. A. Henzinger, P.-H. Ho, and H. Wong-Toi. A user guide to HYTECH. In *Tools and Algorithms for the Construction and Analysis of Systems* (E. Brinksma, W. R. Cleaveland, and K. G. Larsen eds.), Lect. Notes Comp. Sci., vol. 1019, Springer, Berlin, 1995, pp. 41–71.

[77] J. P. Hespanha and A. S. Morse. Stabilization of nonholonomic integrators via logic-based switching. In *Proc. 13th IFAC World Congress, Vol. E: Nonlinear Systems I*, 1996, pp. 467–472.

[78] J. P. Hespanha and A. S. Morse. Stabilization of nonholonomic integrators via logic-based switching. *Automatica*, 35:385–393, 1999.

[79] S.-T. Hu. *Elements of General Topology.* Holden-Day, San Francisco, 1964.

[80] J.-I. Imura and A. J. van der Schaft. *Characterization of well-posedness of piecewise linear systems.* Fac. Math. Sci., Univ. Twente, Memorandum 1475, 1998.

[81] J.-I. Imura and A. J. van der Schaft. Well-posedness of a class of piecewise linear systems with no jumps. In *Hybrid Systems: Computation and Control* (F. W. Vaandrager and J. H. van Schuppen eds.), Lect. Notes Comp. Sci., vol. 1569, Springer, Berlin, 1999, pp. 123–137.

[82] K. H. Johansson, M. Egerstedt, J. Lygeros, and S. Sastry. *Regularization of Zeno hybrid automata.* Manuscript, October 22, 1998. Submitted for publication.

[83] M. Johansson and A. Rantzer. Computation of piecewise quadratic Lyapunov functions for hybrid systems. *IEEE Trans. Automat. Contr.*, 43:555–559, 1998.

[84] T. Kailath. *Linear Systems.* Prentice-Hall, Englewood Cliffs, N. J., 1980.

[85] I. Kaneko and J. S. Pang. Some n by dn linear complementarity problems. *Lin. Alg. Appl.*, 34:297–319, 1980.

[86] C. W. Kilmister and J. E. Reeve. *Rational Mechanics.* Longmans, London, 1966.

[87] S. Kowalewski, S. Engell, J. Preussig, and O. Stursberg. Verification of logic controllers for continuous plants using timed condition/event-system models. *Automatica*, 35:505–518, 1999.

[88] S. Kowalewski, M. Fritz, H. Graf, J. Preussig, S. Simon, O. Stursberg, and H. Treseler. A case study in tool-aided analysis of discretely controlled continuous systems: the two tanks problem. In *Final Program and Abstracts HS'97 (Hybrid Systems V)*, Univ. Notre Dame, 1997, pp. 232–239.

[89] M. Kuijper. *First-Order Representations of Linear Systems.* Birkhäuser, Boston, 1994.

[90] I. Kupka. The ubiquity of Fuller's phenomenon. In *Nonlinear Controllability and Optimal Control* (H. J. Sussmann ed.), Monographs Textbooks Pure Appl. Math., vol. 133, Marcel Dekker, New York, 1990, pp. 313–350.

[91] R. P. Kurshan. *Computer-Aided Verification of Coordinated Processes.* Princeton Univ. Press, Princeton, 1994.

[92] C. Lanczos. *The Variational Principles of Mechanics.* Univ. Toronto Press, Toronto, 1949.

[93] D. M. W. Leenaerts and W. M. G. van Bokhoven. *Piecewise Linear Modeling and Analysis.* Kluwer, Dordrecht, the Netherlands, 1998.

[94] B. Lennartson, M. Tittus, B. Egardt, and S. Pettersson. Hybrid systems in process control. *IEEE Control Systems Magazine*, 16(5):45–56, Oct. 1996.

[95] Y. J. Lootsma, A. J. van der Schaft, and M. K. Çamlıbel. Uniqueness of solutions of relay systems. *Automatica*, 35:467–478, 1999.

[96] P. Lötstedt. Mechanical systems of rigid bodies subject to unilateral constraints. *SIAM J. Appl. Math.*, 42:281–296, 1982.

[97] P. Lötstedt. Numerical simulation of time-dependent contact and friction problems in rigid body mechanics. *SIAM J. Sci. Stat. Comp.*, 5:370–393, 1984.

[98] J. Lygeros, D. N. Godbole, and S. Sastry. Verified hybrid controllers for automated vehicles. *IEEE Trans. Automat. Contr.*, 43:522–539, 1998.

[99] J. Lygeros, C. Tomlin, and S. Sastry. Controllers for reachability specifications for hybrid systems. *Automatica*, 35:349–370, 1999.

[100] O. Maler, ed. *Hybrid and Real Time Systems*. Lect. Notes Comp. Sci., vol. 1386, Springer, Berlin, 1997.

[101] O. Maler. A unified approach for studying discrete and continuous dynamical systems. In *Proc. 37th IEEE Conf. Dec. Contr.*, Tampa, USA, Dec. 1998, pp. 37–42.

[102] O. Maler, A. Pnueli, and J. Sifakis. On the synthesis of discrete controllers for timed systems. In *Theoretical Aspects of Computer Science*, Lect. Notes Comp. Sci., vol. 900, Springer, Berlin, 1995, pp. 229–242.

[103] J. Malmborg, B. Bernhardsson, and K.J. Åstrom. A stabilizing switching scheme for multi controller systems. In *Proc. 13th IFAC World Congress, Vol. F*, 1996, pp. 229–234.

[104] Z. Manna, A. Pnueli. *The Temporal Logic of Reactive and Concurrent Systems: Specification*. Springer, Berlin, 1992.

[105] Z. Manna, A. Pnueli. *Temporal Verification of Reactive Systems: Safety*. Springer, New York, 1995.

[106] C. Marchal. Chattering arcs and chattering controls. *J. Optim. Th. Appl.*, 11:441–468, 1973.

[107] M. Mariton. Almost sure and moments stability of jump linear systems. *Syst. Contr. Lett.*, 11:393–397, 1988.

[108] M. Mariton. *Jump Linear Systems in Automatic Control*. Marcel Dekker, New York, 1990.

[109] S. E. Mattson. On object-oriented modelling of relays and sliding mode behaviour. In *Proc. 1996 IFAC World Congress, Vol. F*, 1996, pp. 259–264.

[110] K. L. McMillan. *Symbolic Model Checking.* Kluwer, Boston, 1993.

[111] J. J. Moreau. Approximation en graphe d'une évolution discontinue. *R.A.I.R.O. Analyse numérique / Numerical Analysis*, 12:75–84, 1978.

[112] J. J. Moreau. Liaisons unilatérales sans frottement et chocs inélastiques. *C. R. Acad. Sc. Paris*, 296:1473–1476, 1983.

[113] J. J. Moreau. Standard inelastic shocks and the dynamics of unilateral constraints. In *Unilateral Problems in Structural Analysis* (G. del Piero and F. Maceri eds.), Springer, New York, 1983, pp. 173–221.

[114] J. J. Moreau. *Numerical aspects of the sweeping process.* Preprint, 1998.

[115] A. S. Morse. Control using logic-based switching. In *Trends in Control* (A. Isidori ed.), Springer, London, 1995, pp. 69–114.

[116] A. S. Morse, C. C. Pantelides, S. S. Sastry, and J. M. Schumacher, guest eds. Special Issue on Hybrid Systems. *Automatica*, 35(3), March 1999.

[117] A. S. Morse and W. M. Wonham. Status of noninteracting control. *IEEE Trans. Automat. Contr.*, 16:568–581, 1971.

[118] R. M. Murray and S. S. Sastry. Nonholonomic motion planning. Steering using sinusoids. *IEEE Trans. Automat. Contr.*, 38:700–716, 1993.

[119] G. M. Nenninger, M. K. Schnabel, and V. G. Krebs. Modellierung, Simulation und Analyse hybrider dynamischer Systeme mit Netz-Zustands-Modellen. *Automatisierungstechnik*, 47:118–126, 1999.

[120] A. Nerode and W. Kohn. Models for hybrid systems: Automata, topologies, controllability, observability. In *Hybrid Systems* (R. L. Grossman, A. Nerode, A. P. Ravn, and H. Rischel eds.), Lect. Notes. Comp. Sci., vol. 736, Springer, New York, 1993, pp. 317–356.

[121] H. Nijmeijer and A. J. van der Schaft. *Nonlinear Dynamical Control Systems.* Springer, New York, 1990.

[122] B. Øksendal. *Stochastic Differential Equations. An Introduction with Applications* (5th ed.). Springer, Berlin, 1998.

[123] E. Ott. *Chaos in Dynamical Systems.* Cambridge Univ. Press, 1993.

[124] L. Paoli, M. Schatzman. Schéma numérique pour un modèle de vibrations avec contraintes unilatérales et perte d'énergie aux impacts, en dimension finie. *C. R. Acad. Sci. Paris, Sér. I Math.*, 317:211–215, 1993.

[125] J. Pérès. *Mécanique Générale*. Masson & Cie., Paris, 1953.

[126] S. Pettersson. *Modeling, Control and Stability Analysis of Hybrid Systems*. Licentiate thesis, Chalmers University, Nov. 1996.

[127] S. Pettersson and B. Lennartson. Stability and robustness for hybrid systems. In *Proc. 35th IEEE Conf. Dec. Contr.*, Kobe, Japan, Dec. 1996, pp. 1202–1207.

[128] S. Pettersson and B. Lennartson. Controller design of hybrid systems. In *Hybrid and Real-Time Systems* (O. Maler ed.), Lect. Notes Comp. Sci., vol. 1201, Springer, Berlin, 1997, pp. 240–254.

[129] F. Pfeiffer and C. Glocker. *Multibody Dynamics with Unilateral Contacts*. Wiley, Chichester, 1996.

[130] V. M. Popov. *Hyperstability of Control Systems*. Springer, Berlin, 1973.

[131] K. Popp and P. Stelter. Stick-slip vibrations and chaos. *Phil. Trans. Roy. Soc. Lond. A*, 332:89–105, 1990.

[132] P. J. Ramadge and W. M. Wonham. Supervisory control of a class of discrete event processes. *SIAM J. Contr. Opt.*, 25:206–230, 1987.

[133] A. W. Roscoe. *The Theory and Practice of Concurrency*. Prentice Hall, London, 1998.

[134] D. Ruelle. *Turbulence, Strange Attractors, and Chaos*. World Scientific, Singapore, 1995.

[135] H. Samelson, R. M. Thrall, and O. Wesler. A partition theorem for Euclidean n-space. *Proc. Amer. Math. Soc.*, 9:805–807, 1958.

[136] S. S. Sastry and C. A. Desoer. Jump behavior of circuits and systems. *IEEE Trans. Circuits Syst.*, 28:1109–1124, 1981.

[137] S. S. Sastry and T. A. Henzinger, eds. *Hybrid Systems: Computation and Control*. Lect. Notes Comp. Sci., vol. 1386, Springer, Berlin, 1998.

[138] A. J. van der Schaft and J. M. Schumacher. The complementary-slackness class of hybrid systems. *Math. Contr. Signals Syst.*, 9:266–301, 1996.

[139] A. J. van der Schaft and J. M. Schumacher. Hybrid systems described by the complementarity formalism. In *Hybrid and Real-Time Systems* (O. Maler ed.), Lect. Notes Comp. Sci., vol. 1201, Springer, Berlin, 1997, pp. 403–408.

[140] A. J. van der Schaft and J. M. Schumacher. Complementarity modeling of hybrid systems. *IEEE Trans. Automat. Contr.*, 43:483–490, 1998.

[141] J. M. Schumacher. Linear systems and discontinuous dynamics. In *Operators, Systems, and Linear Algebra* (U. Helmke, D. Prätzel-Wolters, and E. Zerz eds.), B. G. Teubner, Stuttgart, 1997, pp. 182–195.

[142] A. Seierstad and K. Sydsaeter. *Optimal Control Theory with Economic Applications.* Advanced Textbooks in Economics, vol. 24, North-Holland, Amsterdam, 1987.

[143] R. Sepulchre, M. Jankovic, and P. V. Kokotovic. *Constructive Nonlinear Control.* Springer, London, 1997.

[144] J.-J. E. Slotine and W. Li. *Applied Nonlinear Control.* Prentice-Hall, Englewood Cliffs, New Jersey, 1991.

[145] D. E. Stewart. Convergence of a time-stepping scheme for rigid-body dynamics and resolution of Painlevé's problem. *Arch. Ration. Mech. Anal.*, 145:215–260, 1998.

[146] D. E. Stewart, J. C. Trinkle. An implicit time-stepping scheme for rigid body dynamics with inelastic collisions and Coulomb friction. *Int. J. Numer. Methods Engrg.*, 39:2673–2691, 1996.

[147] H. J. Sussmann. A nonsmooth hybrid maximum principle. In *Proc. Workshop on Nonlinear Control* (Ghent, March 1999), Lect. Notes Contr. Inform. Sci., Springer, Berlin, 1999.

[148] R. Sznajder and M. S. Gowda. Generalizations of p_0- and p-properties; extended vertical and horizontal linear complementarity problems. *Lin. Alg. Appl.*, 223/224:695–715, 1995.

[149] J. G. Thistle and W. M. Wonham. Control of infinite behavior of finite automata. *SIAM J. Contr. Opt.*, 32:1075–1097, 1994.

[150] Ya. Z. Tsypkin. *Relay Control Systems.* Cambridge Univ. Press, Cambridge, 1984.

[151] V. I. Utkin. Variable structure systems with sliding modes. *IEEE Trans. Autom. Contr.*, 22:212-222, 1977.

[152] V. I. Utkin. *Sliding Modes in Control Optimization.* Springer, Berlin, 1992.

[153] F. W. Vaandrager and J. H. van Schuppen, eds. *Hybrid Systems: Computation and Control.* Lect. Notes Comp. Sci., vol. 1569, Springer, Berlin, 1999.

[154] M. Wicks, P. Peleties, and R. DeCarlo. Switched controller synthesis for the quadratic stabilisation of a pair of unstable linear systems. *Eur. J. Contr.*, 4:140–147, 1998.

[155] J. C. Willems. Dissipative dynamical systems. Part I: General theory. *Arch. Rational Mech. Anal.*, 45:321–351, 1972.

[156] J. C. Willems. Paradigms and puzzles in the theory of dynamical systems. *IEEE Trans. Automat. Contr.*, 36:259–294, 1991.

[157] H. P. Williams. *Model Building in Mathematical Programming* (3rd ed.). Wiley, New York, 1993.

[158] P. Wilmott, S. D. Howison, and J. N. Dewynne. *The Mathematics of Financial Derivatives: A Student Introduction*. Cambridge Univ. Press, Cambridge, 1995.

[159] H. S. Witsenhausen. A class of hybrid-state continuous-time dynamic systems. *IEEE Trans. Automat. Contr.*, 11:161–167, 1966.

[160] W. M. Wonham. *Linear Multivariable Control: A Geometric Approach* (3rd ed.). Springer, New York, 1985.

[161] H. Ye, A. N. Michel, and L. Hou. Stability of hybrid dynamical systems. In *Proc. 34th IEEE Conf. Dec. Contr.*, New Orleans, Dec. 1995, pp. 2679–2684.

[162] H. Ye, A. N. Michel, and L. Hou. Stability theory for hybrid dynamical systems. *IEEE Trans. Automat. Contr.*, 43:461–474, 1998.

[163] A. Zavola-Rio and B. Brogliato. Hybrid feedback strategies for the control of juggling robots. In *Modelling and Control of Mechanical Systems* (A. Astolfi *et al.* eds.), Imperial College Press, 1997, pp. 235–252.

Index